T3-BNT-885

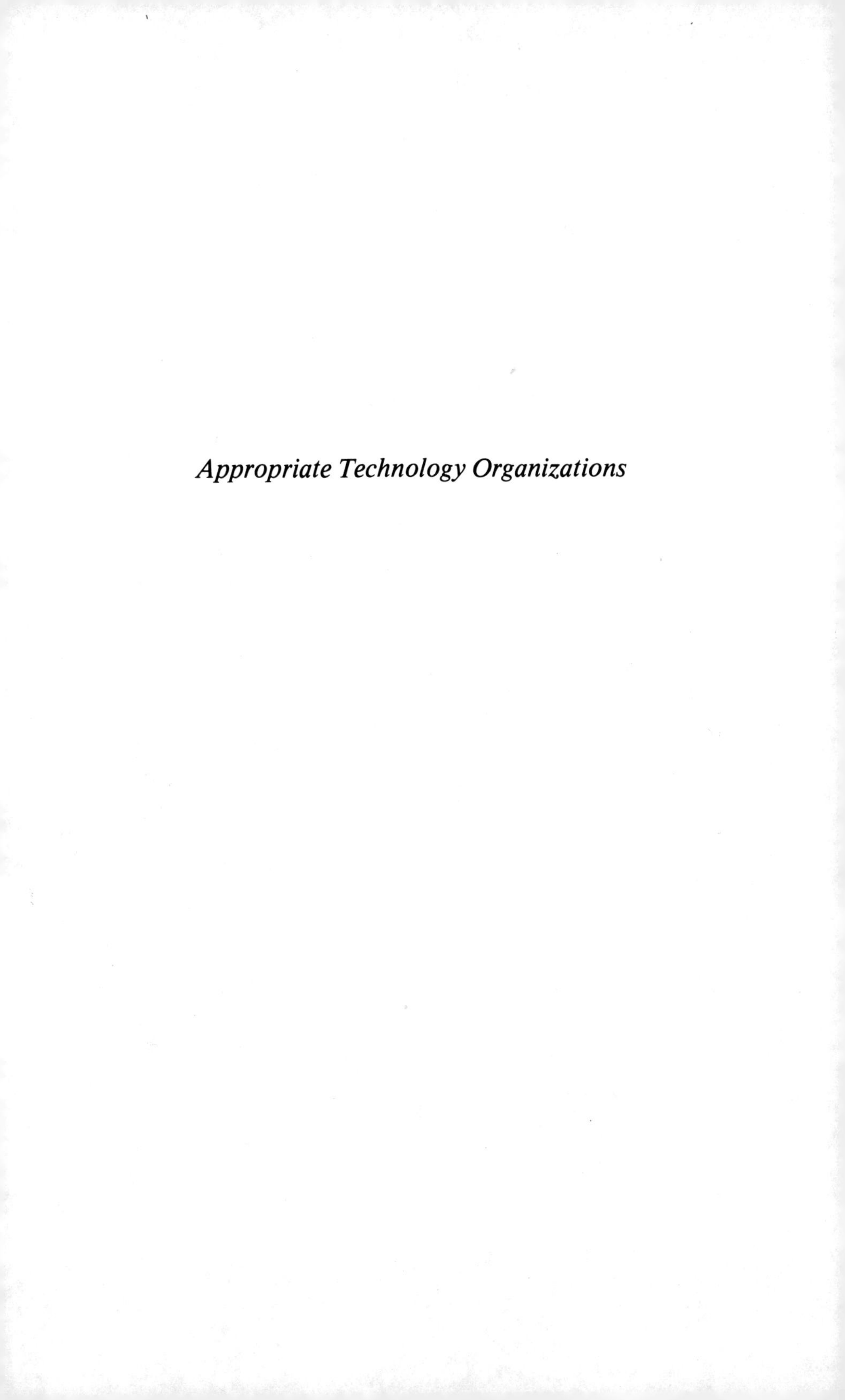

Appropriate Technology Organizations

APPROPRIATE TECHNOLOGY ORGANIZATIONS

A Worldwide Directory

Compiled by the

Center for Business Information
New York, N.Y.

McFarland & Company, Inc., Publishers
Jefferson, N.C., and London

Library of Congress Cataloging in Publication Data

Main entry under title:

Appropriate technology organizations.

Includes index.
1. Appropriate technology—Directories. I. Center for Business Information (U.S.)
T49.5.A657 1984 338.9'025 83-43053

ISBN 0-89950-098-6 (pbk.)

Manufactured in the United States of America

McFarland & Company, Inc., Publishers
Box 611, Jefferson, North Carolina 28640

TABLE OF CONTENTS

INTRODUCTION

The choice of technologies is one of the most important decisions facing a developing country. It is a choice which affects the whole pattern of income distribution and the fabric of the economic and social structure. It determines who works and who does not; where work is done and therefore the urban/rural balance; what is produced; and for whose benefit resources are used.

But developing countries are frequently unaware of the choices open to them. At present the scales are heavily weighted in favor of capital-intensive methods. The bulk of the world's research and development effort is under the control of the rich countries and is devoted to making their technologies increasingly labor-saving and sophisticated. There is a great deal of information about super-technologies, as well as high-pressure salesmanship, both politicial and commercial, behind them.

In contrast, there is no major commercial or political interest in making known the existence of capital-saving and labor-using methods of production. Nor do aid administrators in donor countries possess the necessary knowledge to be able to assist effectively. It is the "appropriate technology" movement which has been filling this gap.

Gradually development agencies are accepting the need for "appropriate technologies," appropriate to situations in developing countries where labor is abundant, where local markets need to be generated and use made of local materials, local skills and organizational capacities. Methods are required which multiply workplaces; that is, which create millions of productive new jobs. This implies technologies which are cheap, simple, small and suitable for the rural areas in which most of the world's population lives. Such jobs should be geared to basic needs in food and food storage, clothing, building materials, education, medicine, water supply and transport. They will also involve agriculturally based industries such as farm equipment manufacture and processing facilities.

The new workplaces should be cheap enough that they can be created in large numbers. If average investment for one workplace is $10,000 (a not unusual figure in industrialized countries), $1,000,000 would provide 100 jobs. At $500 per workplace (a not unattainable average for some technologies), the same money could provide 2,000 jobs, perhaps at a lower productivity per worker but with at least the same, if not greater, total output. The overall gains will be enhanced if the technology mainly makes use of local materials and skills.

If production methods are relatively simple, they will probably be more appropriate to locally available labor skills and capacities for management, organization, and maintenance. Also, they will be less exacting in their requirements as to raw materials in terms of purity or precise specifications.

In many cases the appropriate technologies will also be "intermediate" in terms of size, and the term "intermediate technology" is often used interchangeably with "appropriate technology." "Economics of scale" depend on such local factors as size of market or purchasing power, transport facilities, marketing know-how, managerial skills, manpower supplies and availability of capital. in the situation typical of many developing countries, the most economic scale of production will often be relatively small by comparison with industrialized nations. An appropriate technology might be intermediate in scale as between the hoe and the tractor, or the panga and the combine harvester, in the same way as a bicycle is intermediate between travel on foot and travel by car, yet by no means less useful or efficient, according to the objective in view.

This directory has attempted to list every organization that considers itself involved in appropriate technology, in as many countries as possible. Because of the small nature of such organizations, we have no doubt missed many, and this directory will be one way of bridging the communications gap between workers in appropriate technology. We have made no attempt to list individuals connected with these organizations—they change far too often to be meaningful, and letters addressed by title will achieve far more effective results.

The uses of the directory are many: exchange of newsletters between groups, sources of information on what is being done in a specific area, markets for manufacturers of equipment suitable for use in appropriate technology development, contacts for importers seeking contracts for locally-made goods, and many others.

We would welcome suggestions and corrections for listing in a possible future edition of this directory, and information should be sent to:

Editor
Appropriate Technology Organizations
Center for Business Information
Box 5474
New York NY 10163 U.S.A.

Please do not expect an acknowledgment.

AFRICA (Subsaharan)

(See also NORTH AFRICA..., page 74)

BENIN (DAHOMEY)

001
Centre d'Etudes et de Promotion des Enterprises Beninoises
B.P. 2022
Cononou
BENIN

BOTSWANA

002
Agricultural Information Service
Ministry of Agriculture
Private Bag 3
Gaborone
BOTSWANA

003
Boikanyo Engineering & Service Station
P.O. Box 102
Molepolole
BOTSWANA

004
Botswana Christian Council
P.O. Box 355
Gaborone
BOTSWANA

005
Botswana Development Corporation
Appropriate Technology Centre
Box 438
Gaborone
BOTSWANA

006
Botswana Enterprises Development Unit
Ministry of Commerce & Industry
PMB 004
Gaborone
BOTSWANA

007
Kgatleng Development Board (Mochudi Brigades)
P.O. Box 208
Mochudi
BOTSWANA

008
Kweneng Rural Development Association
Private Bag 7
Molepolole
BOTSWANA

009
Lentswe La Odi Weavers
c/o CUSO
Box 252
Gaborone
BOTSWANA

010
Ministry of Agriculture Women's Extension
Division of Agricultural Extension
Private Bag 3
Gaborone
BOTSWANA

011
Mochudi Farmers Brigade
Kgatleng Development Board
P.O. Box 208
Mochudi
BOTSWANA

012
National Institute for Research in Development and African Studies

(Documentation Unit)
Bag 22
Gaborone
BOTSWANA

013
Pelagano Village Industries
P.O. Box 464
Gaborone
BOTSWANA

014
Rural Industries Innovation Centre
c/o Friederich-Ebert Foundation
P.O. Box 18
Gaborone
BOTSWANA

015
Serowe Brigades Development Trust
P.O. Box 121
Serowe
BOTSWANA

016
Serowe Farmers Brigade
Private Bag 5
Serowe
BOTSWANA

017
Swaneng Hill School
Serowe
BOTSWANA

BURUNDI

018
Eglise Protestante Episcopaleane de Burundi
Buye
B.P. 58
Ngoze
BURUNDI

019
Geoff Bishop
Hospital de Buhiga
D.S. 127
Bujumbura
BURUNDI

020
Gabrielle de Marneffe
B.P. 2520
Bujumbura (via Inades)
BURUNDI

021
I.N.A.D.E.S.
B.P. 2520
Bujumbura
BURUNDI

CAMEROON

022
Association Suisse d'Assistance Technique (SATEC)
B.P. 32
Buea
CAMEROON

023
Centre of Applied Research Pan-African Institute for Development
P.O. Box 133
Buea
CAMEROON

024
Commission pour le Developpement
Federation des Eglises et Missions Evangeliques du Cameroon (FEMEC)
B.P. 790
Yaounde
CAMEROON

025
I.N.A.D.E.S.
B.P. 5
Douala
CAMEROON

026
Institut Pedagogoi Appliquee a Vocation Rurale
B.P. 8
Buea
CAMEROON

027
Oxfam
s/c B.P. 547
Yaounde
CAMEROON

028
Pan African Institute for Development-Central Africa (IPD-AC)
B.P. 4078
Douala
CAMEROON

029
Pan African Institute for Development
The Director
P.O. Box 133
Buea, S.W. Province
CAMEROON

CENTRAL AFRICAN EMPIRE

030
Inter-African Bureau for Soils
OAU/STRC
B.P. 1352
Banqui
CENTRAL AFRICAN EMPIRE

CHAD

031
Bureau de Promotion Industrielle du Chad
B.P. 478
N'Djamena
CHAD

032
Bureau de Promotion Industrielle du Chad
B.P. 458
N'Djamena
CHAD

033
Gilbert Klopfenstein
B.P. 821
N'Djamena
CHAD

034
Projet Karite
B.P. 1116
Sarh
CHAD

035
Societe Chretienne Evangelique du Chad
Travail Rural Chretien
B.P. 801
N'Djamena
CHAD

CONGO

036
Counseil National de la Recherch Scientifique et Technique Ecole Superieure des Sciences
Brazzaville
CONGO

037
OMS, Bureau Regional de l'Afrique
B.P. 6
Brazzaville
CONGO

038
Projet de Developpement Rural
Bureau Internationale du Travail
Nations Unies
B.P. 465
Brazzaville
CONGO

039
Universite Marien Ngouabi
B.P. 69
Brazzaville
CONGO

ETHIOPIA

040
African Rural Housing Association
P.O. Box 2240
Addis Ababa
ETHIOPIA

041
African Training and Research Centre for Women (ATRCW), (UNECA)
P.O. Box 3001
Addis Ababa
ETHIOPIA

042
Appropriate Technology for Farmers
P.O. Box 2003
Addis Ababa
ETHIOPIA

043
Asere Hawariat School
P.O. Box 21495
Addis Ababa
ETHIOPIA

044
Association for the Advancement of Agricultural Sciences in Africa
P.O. Box 30087
Addis Ababa
ETHIOPIA

045
Baptist Mission
Mehal Meda
Shoa, Menz
ETHIOPIA

046
British Transport & Road Research Lab
Box 40125
Addis Ababa
ETHIOPIA

047
Chilalo Agricultural Development Unit (CADU)
P.O. Box 3376
Addis Ababa
ETHIOPIA

048
Camille de Stoop
B.P. 3406 (via INADES)
Soddo-Wolhamo
ETHIOPIA

049
Ethiopian Orthodox Church Development Commission
P.O. Box 503
Addis Ababa
ETHIOPIA

050
Ethiopian Technological Pioneers Centre (ETPC)
The Technical School
Addis Ababa
ETHIOPIA

051
Carl & Vera Hanse
Box 102
Dire Dawa
ETHIOPIA

052
Intermediate Technology Centre
P.O. Box 10417
Addis Ababa
ETHIOPIA

053
Institute of Agricultural Research
P.O. Box 2003
Addis Ababa
ETHIOPIA

054
Institute of Agricultural Research
P.O. Box 103
Nazareth
ETHIOPIA

055
Institute of Development Research
P.O. Box 1176
Addis Ababa
ETHIOPIA

056
International Livestock Centre for Africa
P.O. Box 5689
Addis Ababa
ETHIOPIA

057
Journal for Social Work Education in Africa
Association for Social Work Education in Africa
P.O. Box 1176
Addis Ababa
ETHIOPIA

058
Ministry of Commerce, Industry and Tourism
Small-Scale and Handicraft Industries
P.O. Box 1769
Addis Ababa
ETHIOPIA

059
National University
Faculty of Technology
P.O. Box 518
Addis Ababa
ETHIOPIA

060
National Water Resources Commission
P.O. Box 1008
Addis Ababa
ETHIOPIA

061
Selekleka Agricultural Implements Project
Tigre
ETHIOPIA

062
Source of Docmentation and Communication for Development

P.O. Box 5788
Addis Ababa
ETHIOPIA

063
TAIDL
P.O. Box 26
Makele
ETHIOPIA

064
U.N. Economic Commission for Africa
Science and Technology Section
P.O. Box 3005
Addis Ababa
ETHIOPIA

065
U.N. Economic Commission for Africa (UNECA)
P.O. Box 3001
Addis Ababa
ETHIOPIA

066
U.N. Economic Commission for Africa (UNECA)
Voluntary Agencies Bureau
P.O. Box 3001
Addis Ababa
ETHIOPIA

067
U.N. Economic Commission for Africa
African Training and Research Center for Women
P.O. Box 3001
Addis Ababa
ETHIOPIA

068
Village Technology Innovation Experiment, VTIE
P.O. Box 1107
Asmara
ETHIOPIA

069
Village Technology Innovation Experiment (VTIE)
P.O. Box 31
Goat Hill
Addis Ababa
ETHIOPIA

070
The Workshop
St. Pauls Hospital
Addis Ababa
ETHIOPIA

071
World Vision of Ethiopia
P.O. Box 3330
Addis Ababa
ETHIOPIA

GAMBIA

072
Community Development Centre
Massembe
Kiang East
L.R. Division
GAMBIA

073
Department of Agriculture
The Quadrangle
Banjul
GAMBIA

074
Bishop Rigal Elisee
P.O. Box 51
Banjul
GAMBIA

075
Gambie Co-operative Union Ltd.
P.O. Box 505
Banjul
GAMBIA

076
Indigenous Enterprises Advisory Scheme
c/o Ministry of Economic Planning & Industrial Development
Marina
Banjul
GAMBIA

077
Methodist Church
P.O. Box 288
Banjul
GAMBIA

078
A. Neilson
Small Enterprises Advisor
Presidents Office
Bathurst
GAMBIA

079
W.E.C.
P.O. Box 86
Banjul
GAMBIA

GHANA

080
Agricultural Engineers Ltd.
Ring Road West, Industrial Area
P.O. Box 3707
Accra
GHANA

081
Ashanti Regional Development Corp.
P.O. Box 38
Kumasi
GHANA

082
Christian Service Committee
Agricultural Consultant
P.O. Box 3262
Accra
GHANA

083
Council for Scientific and Industrial Research
Building and Road Research Institute (BRRI)
University P.O. Box 40
Kumasi
GHANA

084
Council for Scientific and Industrial Research
Institute of Standards and Industrial Research
P.O. Box M 32
Accra
GHANA

085
Council for Scientific and Industrial Research
Crops and Research Institute
Kumasi
GHANA

086
Evangelical Presbyterian Church
Social Centre
P.O. Box 224
Ho, Volta Region
GHANA

087
Forest Products Research Institute
University P.O. Box 63
Kumasi
GHANA

088
Ghana Rural Reconstruction Movement
P.O. Box 2338
Accra
GHANA

089
Ghana Science Association
P.O. Box 7
Legon
GHANA

090
Ghana Water & Sewerage Corporation
P.O. Box 24
Bolgatanga
GHANA

091
Industrial Research Institute (IRI)
P.O. Box M 32
Accra
GHANA

092
Moise Mensah
Advisory Committee on Food Production
P.O. Box 1628
Accra
GHANA

093
Ministry of Industries
P.O. Box M 39
Accra
GHANA

094
National Council on Women & Development
Court of Appeal
P.O. Box 119
Accra
GHANA

095
Operation Help Nima
P.O. Box 37

Nima
GHANA

096
Technoserve Inc.
P.O. Box 3262
Accra
GHANA

097
University of Ghana
Faculty of Agriculture
Legon
GHANA

098
University of Ghana
Faculty of Architecture
Legon
GHANA

099
University of Ghana
Department of Nutrition
Legon
GHANA

100
University of Legon
Department of Geography
Accra
GHANA

101
University of Science and Technology
Department of Agricultural Economics
and Farm Management
Kumasi
GHANA

102
University of Science and Technology
Centre for Research & Development
in Housing, Planning, & Building
Faculty of Architecture
Kumasi
GHANA

103
University of Science and Technology
Faculty of Art
Kumasi
GHANA

104
University of Science and Technology
Faculty of Engineering
Kumasi
GHANA

105
University of Science and Technology
Department of Industrial Art
College of Art
Kumasi
GHANA

106
University of Science and Technology
Department of Planning
Kumasi
GHANA

107
University of Science and Technology
Technology Consultancy Centre
Kumasi
GHANA

108
University of Science and Technology
Technology Consultancy Centre—TCC
University Post Office
Kumasi
GHANA

GUINEA

109
Kerfala Cisse Mendemory
B.P. 309
Konakry
GUINEA

IVORY COAST

110
Banque Africaine de Developpement
B.P. 1387
Abidjan
IVORY COAST

111
Compagnie Industrielle d'Amenagement
(CINAM-Abidjan)
B.P. 8389
Abidjan
IVORY COAST

112
Direction des Amenagements Ruraux
B.P.V. 9
Abidjan
IVORY COAST

113
Hospital Protestant

B.P. 115
Dabou
IVORY COAST

114
Institut Africaine pour le Developpement Economique et Social, (I.N.A.D.E.S.)
15, Avenue Jean-Mermoz
Cocody
B.P. 8008
Abidjan
IVORY COAST

115
Institut pour la Technologie et l'Industrialisation des Produits Agricoles Tropicales
B.P. 4549
Abidjan
IVORY COAST

116
Jean-Paul Nicolau
B.P.V. 79
Abidjan
IVORY COAST

117
Office National de la Promotion Rural
B.P. 20225
Abidjan
IVORY COAST

118
Programme d'Aide aux Enterprises Africaines
La Residence Nogues
B.P. 20824
Abidjan
IVORY COAST

119
Societe Ivoirienne d'Expansion Touristique et Hoteliere
B.P. 4375
Abidjan
IVORY COAST

120
UNICEF
West AFrican Regional Office
B.P. 4443
Abidjan
IVORY COAST

KENYA

121
African Co-operative Savings & Credit Association (ACOSCA)
P.O. Box 43278
Nairobi
KENYA

122
Africa Committee for Rehabilitation of Southern Sudan (ACROSS)
P.O. Box 21033
Nairobi
KENYA

123
African Medical & Research Foundation
P.O. Box 30125
Nairobi
KENYA

124
Appropriate Technology Group
RF. Logan-Reid
P.O. Box 454
Kisumu
W. KENYA

125
Bukura Farm Training Centre
P.O. Box 92
Bukura
Kakamega
KENYA

126
Bungoma Farm Training Centre
P.O. Box 46
Bungoma, Western Province
KENYA

127
Canadian International Development Research Centre
P.O. Box 30677
Nairobi
KENYA

128
Christian Industrial Training Centre (CITC)
Box 12935
Pumwani
Nairobi
KENYA

129
Christian Organisations Research & Advisory Trust (CORAT)
P.O. Box 40647
Imani House, St. John's Gate
Nairobi
KENYA

130
Church House and Limuru Conference and Training Centre
Government Road
Nairobi
KENYA

131
Coffee Research Station
Research Liaison Officer
P.O. Box 4
Ruiru
KENYA

132
Jim Collins
(UNDP)
P.O. Box 30513
Nairobi
KENYA

133
East Africa Agricultural and Forestry Research Organisation (EAAFRO)
Agricultural Machinery Coordinating Office
P.O. Box 30148
Nairobi
KENYA

134
East African Industrial Research Organisation
P.O. Box 30650
Nairobi
KENYA

135
E.A. Natural Resources Research Council
P.O. Box 30005
Nairobi
KENYA

136
Egerton College of Agriculture
Librarian
P.O. Njoro
KENYA

137
Extelcoms House
Haile Selassie Avenue
Nairobi
KENYA

138
Farm Machinery Testing Unit
Ministry of Agriculture
Nakuru
KENYA

139
Frederick Ebert Stiftung Foundation
P.O. Box 48413
Nairobi
KENYA

140
Housing Research and Development Unit
University of Nairobi
Box 30197
Nairobi
KENYA

141
Ingotse Christian Training Centre
Church of God, Kenya
P.O. Box 413
Kakamega
KENYA

142
Institute for Development Studies
University of Nairobi
P.O. Box 30197
Nairobi
KENYA

143
International Bank for Reconstruction & Development
P.O. Box 30577
Nairobi
KENYA

144
International Laboratory for Research on Animal Diseases
P.O. Box 47543
Nairobi
KENYA

145
Kaaga Rural Training Centre
P.O. Box 267

Meru
KENYA

146
Kenya Freedom from Hunger Council
P.O. Box 30762
Nairobi
KENYA

147
Kenya Industrial Estates, Ltd.
Rural Industrial Development Program
P.O. Box 18282 Likoni Road
Nairobi
KENYA

148
Kenya Industrial Estates
Product Development Centre
Box 272
Machakos
KENYA

149
Kenya Machinery Testing Unit
P.O. Box 153
Nakuru
KENYA

150
Kinna and Rapsu Irrigation Schemes
P.O. Box 596
Meru
KENYA

151
Lugari Extension Programme
P.O. Box 24
Soy
KENYA

152
Maasai Rural Training Centre
P.O. Box 24
Kajiado
KENYA

153
Maryknoll Fathers
P.O. Box 43058
Nairobi
KENYA

154
Maseno Village Polytechnic
Maseno
KENYA

155
Ministry of Agriculture
P.O. Box 30028
Nairobi
KENYA

156
Ministry of Commerce & Industry
P.O. Box 30430
Nairobi
KENYA

157
Ministry of Commerce & Industry
Industrial Survey & Promotion Centre
Nairobi
KENYA

158
Ministry of Housing & Social Services
Gill House
P.O. Box 30276
Nairobi
KENYA

159
Ministry of Labour
Kenya National Youth Service
P.O. Box 30397
Nairobi
KENYA

160
National Christian Council of Kenya
Church House
Government Road
Box 45009
Nairobi
KENYA

161
Now Build
P.O. Box 20360
Nairobi
KENYA

162
Partnership for Productivity (American)
P.O. Box 243
Kakamega
KENYA

163
Mr. P.D. Paterson
P.O. Box 24909
Nairobi
KENYA

164
Reliance Engineering Works Ltd.
P.O. Box 197
Kisumu
KENYA

165
Rural Development Program
John Boocock
Box 90122
Mombasa
KENYA

166
Rural Training Centre
Maramanti
Meru
KENYA

167
St. Francis Mission Soy
P.O. Box 24
Soy
KENYA

168
St. Michael's Catholic Church
Centre for Bomet Self-help Group
P.O. Box 3009
Bomet
KENYA

169
Samburu Rural Development Centre
P.O. Maralel
via Thompson's Falls
KENYA

170
Timber Industrial Project
Box 30513
Nairobi
KENYA

171
Tunnel Cox Ltd.
Tunnel Estate
Fort Ternana
KENYA

172
Medak Trust
Technical Adviser
P.O. Box 47596
Nairobi
KENYA

173
Ukamba Agricultural Institute
Akamba Hall
P.O. Box 30627
Nairobi
KENYA

174
U.N. Children's Fund
Food Technology & Nutrition Section
Phoenix House, Kenyatta Ave.
P.O. Box 44145
Nairobi
KENYA

175
UNESCO Regional Office of Science
& Technology for Africa
P.O. Box 30592
Nairobi
KENYA

176
UNDP
P.O. Box 30218
Nairobi
KENYA

177
UNICEF
P.O. Box 44145
Nairobi
KENYA

178
UNITERRA
P.O. Box 30552
Nairobi
KENYA

179
University of Nairobi
Department of Agricultural Engineering
P.O. Box 30197
Nairobi
KENYA

180
University of Nairobi
Department of Mechanical Engineering
P.O. Box 30197
Nairobi
KENYA

181
The University of Nairobi
Department of Zoology
P.O. Box 30197

Nairobi
KENYA

182
University of Nairobi
Department of Civil Engineering
P.O. Box 30197
Nairobi
KENYA

183
Village Polytechnics
P.O. Box 45009
Nairobi
KENYA

184
Village Technology Unit
c/o Youth Development Programme
Department of Social Services
Ministry of Housing and Social Services
P.O. Box 30276
Nairobi
KENYA

185
Village Technology Unit
c/o UNICEF
Box 44145
Nairobi
KENYA

LESOTHO

186
BEDCO
Community Development Officer
Teyateyaneng
Box 1216
Maseru
LESOTHO

187
CARE Lesotho
P.O. Box 682
Maseru
LESOTHO

188
Lesotho Agricultural College
Village Technology Development Unit
P.O. Box 829
Maseru
LESOTHO

189
Thaba Bosiu Rural Development Project
Private Bag
Maseru
LESOTHO

190
Thaba Khupa Ecumenical Centre
Farm Institute
Oxen Cultivation Study Group
P.O. Box 929
Maseru
LESOTHO

191
Thaba Khupa Farm Institute
Thaba Khupa Ecomenical Centre
P.O. Box 929
Maseru
LESOTHO

LIBERIA

192
E.N.I. Mission
P.O. Box 167
Greenville
Sinoe County
LIBERIA

193
Department of Home Economics
Ministry of Agriculture
Monrovia
LIBERIA

194
Partnership for Productivity Foundation
LAMCO YEPEKA
Roberts International Airport
LIBERIA

MADAGASCAR

195
Centre d'Information Technique et Economique
Avenue Clemenceau (via de Reviers)
B.P. 74
Malagasy Republic
MADAGASCAR

196
M. Chavanes, B.

SOMALA (via de Reviers)
Ambongalava
par Amparafaravola
Lac Alaotra
MADAGASCAR

197
Church World Service
B.P. 3711
Tananarive
Malagasy Republic
MADAGASCAR

198
Ecole Technique des Freres de Saint Gabriel
ATTN: Frere Dominique (via de Reviers)
Majunga
MADAGASCAR

199
Establissement d'Enseignement
Superieur Polytechnique
Antananarivo
MADAGASCAR

200
M. Laurent
I.E.M.V.T. (via de Reviers)
B.P. 4
Tananarive
MADAGASCAR

MALAWI

201
Agricultural Research Station
Farm Machinery Testing Unit
Chitedze
P.O. Box 158
Lilongwe
MALAWI

202
Chancellor College
Department of Chemistry
P.O. Box 280
Zomba
MALAWI

203
Christian Service Committee of the Churches in Malawi
P.O. Box 949
Blantyre
MALAWI

204
Department of Forestry
Divisional Forest Officer (Wood Energy)
P.O. Box 30048
Capital City
Lilongwe 3
MALAWI

205
Diocese of Southern Malawi
Engineering Department
P/A Malindi
P.O. Mongochi
MALAWI

206
Lilongwe Land Development Programme
Lilongwe
MALAWI

207
Malawi Polytechnic
Private Bag 303
Blantyre 3
MALAWI

208
Rural Trade School
P.O. Box 34
Salima
MALAWI

209
Oxford Committee for Famine Relief (OXFAM)
Field Director
P.O. Box 1363
Blantyre
MALAWI

MALI

210
Compagnie Malienne des Textiles (CMDT)
B.P. 487
Bamako
MALI

211
Direction du Materiel Agricole
B.P. 155
Bamako
MALI

212
Ecole Nationale d'Ingenieurs

B.P. 119
Bamako
MALI

213
Ecole Normale Superieure
Centre de Recherche pour l'Utilisation de l'Energie Solaire
B.P. 134
Bamako
MALI

214
Institut d'Economie Rurale
Ministere du Developpement Rural
Bamako
MALI

215
Institut Polytechnique Rural
Katibougou
MALI

216
Dr. Pat Kelly
B.P. 105
rue 34, angle 33
Quinzembougou
Bamako
MALI

217
Laboratoire de l'Energie Solaire du Mali
B.P. 134
Bamako
MALI

218
SMECMA
Bamako
MALI

MAURITIUS

219
Biscuits de Manioc
Mahebourg
MAURITIUS

220
Mauritius Sugar Industry Research Institute
Reduit
MAURITIUS

221
University of Mauritius
Reduit
MAURITIUS

222
University of Mauritius
School of Industrial Technology
Reduit
MAURITIUS

223
University of Mauritius
School of Agriculture & Industrial Technology
Reduit
MAURITIUS

MOZAMBIQUE

224
Acrow Boror
C.P. 79
Matola
MOZAMBIQUE

225
Universidad de Maputo
Curso de Engenharia Electrotecnica
C.P. 257
Maputo
MOZAMBIQUE

226
Universidad de Maputo
Departamento de Engenharia Quimica
Av. de Mocambique
KM 2
Maputo
MOZAMBIQUE

NIGER

227
Lutheran World Relief
ATTN: Gary Eilerts
B.P. 11624
Niamey
NIGER

228
Church World Service
B.P. 624
Niamey
NIGER

229
Euro-Action Sahel
B.P. 624
Niamey
NIGER

NIGERIA

230
A.A. Afonja
P.O. Box 1912
Ibadan
NIGERIA

231
African Rural Storage Centre
IITA
PMB 5320
Ibadan
NIGERIA

232
Ahmadu Bello University
Department of Agricultural Economics
Institute for Agricultural Research
PMB 1044
Samaru, Zaria
NIGERIA

233
Ahmadu Bello Univeristy
Department of Civil Engineering
Zaria
NIGEF.IA

234
Ahmadu Bello University
Extension & Research Liaison Service
Institute for Agricultural Research
PMB 1044
Zaria
North Central State
NIGERIA

235
Ahmadu Bello University
Rural Economy Research Unit (RERU)
Institute for Agricultural Research
Samaru
PMB 1044
Zaria
NIGERIA

236
Christian Council of Nigeria
P.O. Box 184
Abeokuta
Western State
NIGERIA

237
Christian Council of Nigeria
East Central State
Committee for Agricultural & Rural Development
P.O. Box 84
Okigwe
E.C. State
NIGERIA

238
Christian Rural Advisory Council
P.O. Box 386
7 Tudun Wada Road
Jos, Benue Plateau State
NIGERIA

239
College of Technology
PMB 1110
Clabar
NIGERIA

240
S.W. Eaves
P.O. Box 401
Zaria
N.C. State
NIGERIA

241
Federal Institute of Industrial Research
P.O. Box 1023
Lkeja, Lagos
NIGERIA

242
Federal Institute of Industrial Resources (FIIR), Oshodi
Office of the Director of Research
Private Mail Bag 1023
Murtala Muhammed Airport, Lagos State
NIGERIA

243
Federal Ministry of Industries
Industrial Development Centre
PMB 1035
Zaria, North Central State
NIGERIA

244
Federal Ministry of Industries
Small Industries Division

Lagos
NIGERIA

245
Ford Foundation
47 Marina
P.O. Box 2368
Lagos
NIGERIA

246
Institute of Church & Society
P.O. Box 4020
Ibadan
NIGERIA

247
Integrated Education for Development (van Leer Fndn)
6 Tudin Wada Road
P.O. Box 2174
Jos, B.P. State
NIGERIA

248
Intermediate Technology Workshops
P.O. Box 401
Zaria, N.C. State
NIGERIA

249
International Institute of Tropical Agriculture (IITA)
PMB 5320
Ibadan
NIGERIA

250
Limb Fitting Centre
Hopeville
Uturu, East Central State
NIGERIA

251
Nigerian Stored Products Research Institute
Harvey Road, Yaba
PMB 12543
Lagos
NIGERIA

252
Rev. John Ockers
B.P. 121
Maradi
NIGERIA

253
Office de l'Energie Solaire (ONERSOL)
B.P. 621
Niamey
NIGERIA

254
Projects Development Agency
3, Independence Layout
P.O. Box 609
Enugu
NIGERIA

255
Rehabilitation Centre
Hopeville
Okigwe
East Central State
NIGERIA

256
Secours Mondial Lutherien
B.P. 624
Niamey
NIGERIA

257
Trainee Workshop
P.O. Box 401
Waff Road
Zaria
N.C. State
NIGERIA

258
University of Ibadan
Dept. of Preventative and Social Medicine
Ibadan
NIGERIA

259
University of Ife
Extension Education & Rural Sociology
Faculty of Agriculture
Ile-Ife
NIGERIA

260
University of Ife
Dept. of Chemistry/Chemical Engineering
Ile-Ife
NIGERIA

261
University of Ife
Dept. of Food Science and Technology
Ile-Ife
NIGERIA

262
University of Ife
Industrial Research Unit
Ile-Ife
NIGERIA

263
University of Lagos
Dept. of Chemical Engineering
Lagos
NIGERIA

264
University of Nigeria
Dept. of Mechanical Engineering
Nsukka
NIGERIA

265
University of Nigeria
Dept. of Physics
Nsukka
NIGERIA

266
Vom Hospital
Faith & Farm
P.O. Vom, via Jos
Benue-Plateau State
NIGERIA

267
The Rev. Martin N. Wright
4 Harley Street
P.O. Box 592
Port Harcourt
NIGERIA

268
Okon Equere and Edet Fkanem
473 Atu Street
P.O. Box 191
Calabar
S.E. State
NIGERIA

RHODESIA-ZIMBABWE

269
Berejena Mission
Me-Wo Training Centre
P. Bag 9069
Fort Victoria
RHODESIA/ZIMBABWE

270
Butchers Shop in the Backyard
Old Umtali Mission
P.O. Box 7024
Umtali
RHODESIA/ZIMBABWE

271
Friends Rural Service Centre
P.O. Box 708
Bulawayo
RHODESIA/ZIMBABWE

272
The Nyafaru Development Co. (Pvt) Ltd.
P.O. Box 24
Troutbeck, via Umtali
RHODESIA/ZIMBABWE

273
The Seminary
Chicawasha
P.O. Box 1139
Salisbury
RHODESIA/ZIMBABWE

RWANDA

274
Association Rwandaise des Compagnons Batisseurs
B.P. 454
Kigali
RWANDA (CENTRAL AFRICA)

275
Centre d'Etudes et d'Applications de l'Energie au Rwanda (CEAER)
Dept. de Physique
Universite Nationale du Rwanda
Butare
RWANDA (CENTRAL AFRICA)

276
Economat General
B.P. 30
Kibungo
RWANDA

277
Eglise Anglicane au Rwanda
B.P. 61
Kigali
RWANDA (CENTRAL AFRICA)

278
Hospital Kirinda

El Ing HTDL
B.P. 67
Gitarama
RWANDA (CENTRAL AFRICA)

279
I.N.A.D.E.S.
B.P. 866
Kigali
RWANDA (CENTRAL AFRICA)

280
l'Ecole Agricole
Butare
RWANDA (CENTRAL AFRICA)

281
Rural Environment and Housing in Inter-Tropical Africa
Butare
RWANDA

282
Rwanda Mission, CMS
Kigeme Hospital
B.P. 43
Gikongoro
RWANDA

SENEGAL

283
BUD Senegal
24 Avenue Roume
B.P. 423
Dakar
SENEGAL

284
Centre de Recherches pour le Developpement International
B.P. 11007
Dakar
SENEGAL

285
Centre Experimental de Recherches et d'Etudes pour l'Equipement (CEREEQ)
B189
Hann
Dakar
SENEGAL

286
CER Min. du Developpement Rural
Administration Building
Fourth Floor
Dakar
SENEGAL

287
Council for the Development of Economic & Social Research in Africa (CODESRIA)

B.P. 3186
Dakar
SENEGAL

288
Direction Generale de la Recherche Scientifique et Technique (DGRST)
61 Bd. Pinet Laprade
Dakar
SENEGAL

289
Ecole Nationale D'Economie Appliquee (ENEA)
Route de Ouakam
B.P. 5084
Dakar
SENEGAL

290
ENDA—Environment Training Programme
Technology Relay for Ecodevelopment and Planning in African Environments
(RETED)
B.P. 3370
Dakar
SENEGAL

291
Famille & Development
B.P. 11.007
C.D. Annexe
66, avenue de la Republique
Dakar
SENEGAL

292
Institut de Technologie Alimentaire
B.P. 2765
Dakar
SENEGAL

293
International Crops Research Institute for the Semi-Arid Tropics (ICRISAT)
28 Rue Thiers

B.P. 3340
Dakar
SENEGAL

294
International Development Research Centre
B.P. 11007
C.D. Annexe
Dakar
SENEGAL

295
I.U.T./Department de Genie Mecanique
Dakar
SENEGAL

296
Maisons Familiales Rurales
B.P. 55
Thies
SENEGAL

297
Mini du Developpement Rural
Secretariat Executif des Centres d'Espansion Rurale
Building Administratif
Dakar
SENEGAL

298
Societe Industrial pour l'Application de l'Energie Solaire (SINAES)
B.P. 1277
Dakar
SENEGAL

299
Societe Industrielle Senegalaise de Construction Mecaniques et de Materiels Agricoles (SISCOMA)
B.P. 3214
Dakar
SENEGAL

300
UN Environment Programme Environment Training
B.P. 3370
Dakar
SENEGAL

301
UNESCO Regional Office for Education in Africa
P.O. Box 3311
Dakar
SENEGAL

302
Universite de Dakar
Faculte des Sciences
Dakar
SENEGAL

SIERRA LEONE

303
Centre for Advisory Services in Technology, Research and Development (ASTRAD)
Dept. of Engineering
Fourah Bay College
University of Sierra Leone
Freetown
SIERRA LEONE

304
Fourah Bay College
Dept. of Engineering
Freetown
SIERRA LEONE

305
Donald Hamilton
Ministry of Social Welfare & Rural Development
Fort Street
Freetown
SIERRA LEONE

306
Njala University College
Dept. of Agricultural Engineering
P.M.B.
Freetown
SIERRA LEONE

307
Tikonko Agricultural Extension Centre
Small Farm Equipment Unit
P.O. Box 142
Bo
SIERRA LEONE

SOUTH AFRICA

308
Africa Institute of S. Africa
Unisa Building
P.O. Box 630
Pretoria
S. AFRICA

309
J.B. Arkin
P.O. Box 23687
Joubert Park
Johannesburg
S. AFRICA

310
Bantu Investment Corporation of S.A., Ltd.
P.O. Box 213
Pretoria
S. AFRICA

311
Council for Scientific & Industrial Research
P.O. Box 395
Pretoria 0001
S. AFRICA

312
Environmental & Development Agency (EDA)
P.O. Box 62054
Marshalltown
2107 Transvaal
S. AFRICA

313
The Evangelical Alliance Relief Fund S.A. (TEAR)
P.O. Box 1565
Capetown
S. AFRICA

314
Rev. Makepeace M. Nomvete
Private Bag 9970
Ladysmith
Natal
S. AFRICA

315
Organic Soil Association of South Africa
Box 47100
Parklands
Johannesburg
SOUTH AFRICA

316
South African Council of Churches
P.O. Box 31190
Braamfontein 2017
Transvaal
SOUTH AFRICA

317
South African Council of Churches
Diakonia House
80 Jorissen St.
Johannesburg 2001
SOUTH AFRICA

318
South African Sugar Association
P.O. Box 507
Durban
SOUTH AFRICA

319
Southern African Technical Development Group (Pvt) Ltd.
P.O. Box 31519
Braamfontein
Johannesburg
SOUTH AFRICA

320
South African Voluntary Service
P.O. Box 97
Johannesburg
SOUTH AFRICA

321
Sugar Milling Research Institute
University of Natal
Francois Road
Durban
SOUTH AFRICA

322
The Valley Trust
P.O. Box 33
Botha's Hill
3660 Natal
SOUTH AFRICA

323
Zululand Churches Health & Welfare Association
Kwanzimela Agricultural Project
Private Bag 802
Melmoth 3835
SOUTH AFRICA

SWAZILAND

324
Ministry of Industry, Mines and Tourism
Small Enterprises Promotion Development Corporation
P.O. Box 451

Mbabane
SWAZILAND

325
National Industrial Development Corporation
P.O. Box 99
Malkerns
SWAZILAND

326
National Industrial Development Corp.
Matsana Industrial Estate
P.O. Box 50
Manzini
SWAZILAND

327
NIDCS Tractor Project
P.O. Box 450
Manzini
SWAZILAND

328
Small Enterprises Development Corp.
P.O. Box 451
Mbabane
SWAZILAND

TANZANIA

329
Arusha Appropriate Technology Pilot Project
P.O. Box 764
Arusha
TANZANIA

330
Behemba Rural Training Centre
P.O. Box 160
Musoma
TANZANIA

331
Catholic Relief Services
P.O. Box 9222
Dar-es-Salaam
TANZANIA

332
Christian Council of Tanzania
P.O. Box 2537
Dar-es-Salaam
TANZANIA

333
Community Development Trust Fund
P.O. Box 9133
Dar-es-Salaam
TANZANIA

334
H.A. Foshbrooke
P.O. Box 1268
Arusha
TANZANIA

335
Kasulu Rural Training Centre
Kigoma
TANZANIA

336
Kibaha Education Centre
Box 30054
Kibaha, DSM
TANZANIA

337
Kilacha Production & Training Centre
P.O. Box 8021
Himo, near Moshi
TANZANIA

338
Liberation Support Group
P.O. Box 2099
DSM
TANZANIA

339
Lushoto Integrated Development Project
P.O. Box 60
Soni
TANZANIA

340
Ministry of Commerce and Works
P.O. Box 9144
Dar-es-Salaam
TANZANIA

341
Ministry of Lands, Housing and Urban Development
P.O. Box 9132
Dar-es-Salaam
TANZANIA

342
National Housing Corp.
P.O. Box 2977
Dar-es-Salaam
TANZANIA

343
Ndoleleji Womens Development Programme
S.L.P. 229
Shinyanga
TANZANIA

344
Operation Bootstrap
Box 556
Arusha
TANZANIA

345
Research & Training Institute
Private Bag
Ilonga
Morogoro Region
TANZANIA

346
Shirati Hospital Workers Co-op Store
Shirati Hospital
Private Bag
Musoma
TANZANIA

347
Small Industries Development Organization (SIDO)
Government of Tanzania
Arusha
TANZANIA

348
SIDO Project
P.O. Box 461
Dodoma
TANZANIA

349
Small Industries Development Organization (SIDO)
P.O. Box 2476
Dar-es-Salaam
TANZANIA

350
Small Industries Division
P.M.'s Office
P.O. Box 980
Dodoma
TANZANIA

351
Baptist Agricultural Project
Box 172
Tukuyu
TANZANIA

352
Tanga Integrated Development Programme
P.O. Box 5047
Tanga
TANZANIA

353
Tanzania Agricultural Machinery Testing Unit
Intermediate Technology Section
P.O. Box 1148
Arusha
TANZANIA

354
Tanzania Agricultural Machinery Testing Unit
P.O. Box 1389
Arusha
TANZANIA

355
Tanzania-Canada Beekeeping Project
P.O. Box 661
Arusha
TANZANIA

356
Tanzania Housing Bank
P.O. Box 1723
Dar-es-Salaam
TANZANIA

357
Tanzania Publishing House
47 Independence Are
P.O. Box 2128
Dar-es-Salaam
TANZANIA

358
Ubungo Farm Implements
P.O. Box 20126
Dar-es-Salaam
TANZANIA

359
UNI DSM FAC AGRIC
Box 643
Morogoro
TANZANIA

360
University of Dar-es-Salaam
Dept. of Agricultural Mechanization
Faculty of Agriculture
P.O. Box 643

Morogoro
TANZANIA

361
University of Dar-es-Salaam
Economic Research Bureau
P.O. Box 35096
Dar-es-Salaam
TANZANIA

362
University of Dar-es-Salaam
Dept. of Electrical Engineering
P.O. Box 24121
Dar-es-Salaam
TANZANIA

363
University of Dar-es-Salaam
Faculty of Engineering
P.O. Box 35131
Dar-es-Salaam
TANZANIA

364
University of Dar-es-Salaam
Institute of Development Studies
P.O. Box 35169
Dar-es-Salaam
TANZANIA

365
University of Dar-es-Salaam
Bureau of Research Assessment & Land Use Planning (BRALUP)
P.O. Box 35097
Dar-es-Salaam
TANZANIA

366
University of Dar-es-Salaam
Dar-es-Salaam
TANZANIA

367
Uyole Agricultural Centre
P.O. Box 400
Mbeya
TANZANIA

368
F.J.H. van de Laak
P.O. Box 168
Shinyanga
TANZANIA

369
Y.M.C.A. Farm School
P.O. Box 985
Moshi
TANZANIA

TOGO

370
Centre de Construction et du Logement
B.P. 1762
Cacavelli
Lome
TOGO

371
Centre National de Promotion de Petites et Moyennes Enterprises
B.P. 1086
Lome
TOGO

372
Corps de la Paix
B.P. 3194
Lome
TOGO

373
Dialogues Internationaux en Afrique Occidentale
B.P. 971
Lome
TOGO

374
Eglise Evangelique du Togo
United Church Board for World Ministries
ATTN: James Winter
B.P. 79
Atakpame
TOGO

375
I.N.A.D.E.S.
B.P. 3 (C.c.p. Inades Lome 01-91)
Dapango
TOGO

376
Ministere du Commerce de l'Industrie
Direction de l'Industrie
B.P. 831
Lome
TOGO

377
Rural Development Consultancy for

Christian Churches in Africa
27 rue Kayigan
B.P. 1857
Kodjoviakope
Lome
TOGO

378
Small-Scale Industries Programme
c/o U.N. Development Programme
B.P. 911
Lome
TOGO

UGANDA

379
Africa Basic Foods Inc.
P.O. Box 3140
Kampala
UGANDA

380
Christian Development Agency
P.O. Box 745
Kampala
UGANDA

381
E.A. Development Bank
P.O. Box 7128
Kampala
UGANDA

382
Karamoja Agricultural project
B.C.M.S.
P.O. Box 44
Moroto
UGANDA

383
Makerere Institute of Social Research
P.O. Box 16022
Kampala
UGANDA

384
Makerere University
Dept. of Chemistry
P.O. Box 7062
Kampala
UGANDA

385
Makerere University
Dept. of Crop Science
P.O. Box 7062
Kampala
UGANDA

386
Makerere University
Faculty of Technology
P.O. Box 7062
Kampala
UGANDA

387
Uganda-Co-operative Alliance Ltd.
47, 49 Nkrumah Road
P.O. Box 2215
Kampala
UGANDA

388
Uganda Development Organisation Ltd.
P.O. Box 7042
Kampala
UGANDA

389
UN Development Programme Office
P.O. Box 7184
Kampala
UGANDA

UPPER VOLTA

390
Africare
Oxfam
B.P. 489
Ouagadougou
UPPER VOLTA

391
Atelier Regional de Confection de Material Agricole
(A.R.Co.M.A.)
Bobo Dioulasso
UPPER VOLTA

392
Catholic Relief Service
B.P. 469
Ouagadougou
UPPER VOLTA

393
Centre d'Application des Technologies Rurales et Urbaines (CATRU)
C.N.P.A.R.

B.P. 575
Ouagadougou
UPPER VOLTA

394
Centre des Artisans Buraux
Bobo Dioulasso
UPPER VOLTA

395
Centre des Artisans Ruraux
B.P. 575
Ouagadougou
UPPER VOLTA

396
Centre National de Perfectionnement des Artisans Ruraux (CNPAR)
B.P. 575
Ouagadougou
UPPER VOLTA

397
Comite de Coordination du Developpement Rural
Ouagadougou
UPPER VOLTA

398
Comite Interafricain d'Etudes Hydrauliques (CIEH)
B.P. 868
Ouagadougou
UPPER VOLTA

399
Comite Inter-etats de Lutte contre la Secheresse dans le Sahel
B.P. 7049
Ouagadougou
UPPER VOLTA

400
Corps de la Paix
B.P. 537
Ouagadougou
UPPER VOLTA

401
Ecole Inter-Etats d'Ingenieurs de l'Equipment Rural (EIER)
B.P. 7023
Ouagadougou
UPPER VOLTA

402
ESSOR Rural
Editor-in-Chief
B.P. 7007
Ouagadougou
UPPER VOLTA

403
Michelle Goldsmit
B.P. 130
CAP Matoukou
Bobo Dioulasso
UPPER VOLTA

404
Inter-Etats des Techniciens Superieur de l'Hydraulique et de Equipement Rural (ETSHER)
B.P. 594
Ouagadougou
UPPER VOLTA

405
Jeunesse Rurale de l'Economie Familiale Nutrition
Ouagadougou
UPPER VOLTA

406
Inter-African Committee for Hydraulic Studies
B.P. 369
Ouagadougou
UPPER VOLTA

407
Mission Catholique Tema-Boken
par Kaya
UPPER VOLTA

408
Mission Protestante
B.P. 128
Bobo Dioulasso
UPPER VOLTA

409
New Ayoma Youth Association
c/o P.O. Box 16
New Ayoma
UPPER VOLTA

410
Office de Promotion de l'Entreprise Voltaique (OPEV)
B.P. 188
Ouagadougou
UPPER VOLTA

411
Oxfam
B.P. 489
Ouagadougou
UPPER VOLTA

412
Programme des Nations Unies pour le Developpement
B.P. 573
Ouagadougou
UPPER VOLTA

413
Projet d'Egalite d'Acces des Femmes et des Jeunes Filles a l'Education
Min. l'Education Nationale
Ouagadougou
UPPER VOLTA

414
Projet Habitat (CISSIN)
PNUD, B.P. 7014
Ouagadougou
UPPER VOLTA

415
SAED Section de Technologie Adaptee
B.P. 489
Ouagadougou
UPPER VOLTA

416
Save the Children Fund
B.P. 489
Ouagadougou
UPPER VOLTA

417
Societe d'Etudes et de Developpement (SAED)
B.P. 593
Ouagadougou
UPPER VOLTA

ZAIRE

418
Association pour le Developpement Integral-Bondaba
Internationale Bouworde, Belgium
B.P. 49
Lisala
ZAIRE

419
Batiment Royal
Office de Promotion de l'Enterprise Zaioise (OPEZ)
Avenue du 30 juin
Kinshasa
ZAIRE

420
Bureau Technique de Coordination des Agents de Developpement
Institute Superieur de Developpement Rural
B.P. 2849
Bukava, Kivu
ZAIRE

421
Centre Communautaire de Developpement
B.P. 70
Kimpese
ZAIRE

422
Centre de Developpement Communautaire (CEDECO)
B.P. 170
Kimpese
Bas-Zaire
ZAIRE

423
Centre d'Etudes pour l'Action Sociale (CEPAS)
9, Avenue Pere Boka
B.P. 3096
Kinshasa, Gombe
ZAIRE

424
Comite de Co-ordination pour la Cooperation au Development en Republique du Zaire (CODESA)
B.P. 7611
Kinshasa 1
ZAIRE

425
Eglise Kimbanguiste
B.P. 11375
Kinshasa
ZAIRE

426
Equatorial Africa Office
Deleque Regional
B.P. 10362
Kinshasa 1
ZAIRE

427
Pere Henri Farcy
B.P. 2849
Bukavu, Kivu
ZAIRE

428
Internationale Bouworde (Belgium)
Association pour le Developpement
Integral-Bondaba
B.P. 49
Lisala
ZAIRE

429
I.N.A.D.E.S.
B.P. 3096
Kinshasa
ZAIRE

430
Service de Developpement Agricole
(SEDA)
Nyanga
B.P. 1
Tshikapa
ZAIRE

431
Service du Vulgarisation
P.A.P.
B.P. 10
Kikwit
ZAIRE

432
Unioncoop
B.P. 187
Mbuji-Mayi
Kasai-Oriental
ZAIRE

433
United Methodist Community
B.P. 226
Lodja
ZAIRE

434
Vanga Hospital
B.P. 4728
Kinshasa II
ZAIRE

435
Fr. Antoine Verwilghon
Panzi
s/c P.B. 7.245
Kinshasa I
ZAIRE

ZAMBIA

436
Chikuni Rural Industries
Canisius Secondary School
P.O. Chisekesi
Southern Province
ZAMBIA

437
Chilonga Hospital
P.O. Box 30
Mpika
ZAMBIA

438
Department of Agriculture
P.O. Box 41
Solwezi
ZAMBIA

439
Family Farms Ltd.
Bos RW 285
Lusaka
ZAMBIA

440
Farm Machinery Research Unit
Regional Research Station
P.O. Box 11
Magoye
ZAMBIA

441
"Farming Voice"
P.O. Box 2130
Lusaka
ZAMBIA

442
Tony Finch
Forest Department
P.O. Box 228
Ndola
ZAMBIA

443
Ministry of Rural Development
Natural Resources Development College
P.O. Box Ch. 99
Lusaka
ZAMBIA

444
Natural Resources Development College
P.O. Box Ch. 99
Lusaka
ZAMBIA

445
Northern Technical College
Mechanical Section
P.O. Box 1583
Ndola
Copperbelt Province
ZAMBIA

446
E.R. Potts
P.O. Box 519
Chipata
ZAMBIA

447
Rucom Industries Ltd.
Longolongo Street
P.O. Box 800
Lusaka
ZAMBIA

448
Social Action in Lusaka (SAIL)
Box 3019
Church Road
Lusaka
ZAMBIA

449
University of Zambia
JETS of Zambia
School of Engineering
P.O. Box 2379
Lusaka
ZAMBIA

450
University of Zambia
Rural Development Studies Bureau
P.O. Box 900
Lusaka
ZAMBIA

451
University of Zambia
ATTN: Manager
Technology Development & Advisory
Unit
P.O. Box 2379
Lusaka
ZAMBIA

452
Zambian Council for Social Development (ZCSD)
P.O. Box RW 369
Lusaka
ZAMBIA

ASIA AND PACIFIC

AUSTRALIA

453
Action for World Development
Getrude Street
Fitzoy, Victoria 3065
AUSTRALIA

454
Alternative Technology Unit (ATU)
Dept. of Agriculture
Sydney University
Sydney
AUSTRALIA

455
Appropriate Technology and Community Environment (APACE)
P.O. Box 770
North Sydney
New South Wales 2060
AUSTRALIA

456
Appropriate Technology Development Group
20 Holdsworth Street
Woollahra
New South Wales 2025
AUSTRALIA

457
Australian Council for Overseas Aid
P.O. Box 1563
Canberra City, ACT 2061
AUSTRALIA

458
Australian Development Assistance Agency
P.O. Box 887
Canberra City 2601
AUSTRALIA

459
Commonwealth Scientific and Industrial Research Organisation (CSIRO)
Box 225
Dickson, A.C.T. 2602
AUSTRALIA

460
Commonwealth Scientific Research Organisation
Graham Road
Highett, Melbourne
Victoria 3190
AUSTRALIA

461
Davey Dunlite Co.
Adelaide
AUSTRALIA

462
Energy Conversion Group
Dept. of Engineering Physics
Research School of Physical Sciences
Australian National University
Canberra
AUSTRALIA

463
The Flinders University
Bedford Par
South AUSTRALIA 5042

464
Freedom from Hunger Campaign
69 Clarence Street
Sydney 2001
AUSTRALIA

465
Friends of the Earth
P.O. Box 5115
Townsville, Queensland
AUSTRALIA

466
Intermediate Technology Development Pty, Ltd.
87, Riawena Road
Rose Bay
Hobart, Tasmania
AUSTRALIA

467
International Solar Energy Society (ISES)
P.O. Box 52
Parkville
Victoria 3052
AUSTRALIA

468
Soil Association of South Australia
6 Bickham Court
Dernancourt
South AUSTRALIA 5075

469
Sydney University Eco-Tech Workshop
Sydney, New South Wales
AUSTRALIA

470
University of Queensland
St. Lucia
AUSTRALIA

471
Walkers Limited
P.O. Box 211
Maryborough, Queensland
AUSTRALIA

472
World Christian Action
100 Flanders Street
Melbourne, Victoria
AUSTRALIA

BANGLADESH

473
ADAB NEWS
549F, Road 14
Dhanmondi, Dacca-5
BANGLADESH

474
Agricultural Development Agencies in Bangladesh
549F, Road 14
Dhanmandi, Box 5945
Dacca-5
BANGLADESH

475
Appropriate Agricultural Technology Cell
Bangladesh Agricultural Research Council
130-C, Road No. 1
Dhanmondi Residential Area
Dacca-5
BANGLADESH

476
Asia Foundation
Representative
15/A, Road 6, Dhanmondi
Dacca-5
BANGLADESH

477
Association of Voluntary Agencies in Bangladesh
P.O. Box 5045
Dacca 5
549F, Road 14
Dhanmandi
Dacca 5
BANGLADESH

478
Australian Baptist Missionary Society
42 Green Road
Dacca
BANGLADESH

479
Bangladesh Academy of Rural Development
Kotbari, Comilla
BANGLADESH

480
Bangladesh Agricultural Research Council
Storage and Handling
Appropriate Technology Cell
130C, Road 1
Dhanmandi, Dacca
BANGLADESH

481
Bangladesh Council for Scientific and Industrial Research
Dacca
BANGLADESH

482
Bangladesh National Scientific and

Technical Documentation Centre
Dacca 5
BANGLADESH

483
Bangladesh Small-Cottage Industry Corporation
137/138 Motijheel Commercial Area
Dacca-2
BANGLADESH

484
Bangladesh Rural Advancement Committee (BRAC)
3 New Circular Road, Moghbazar
Dacca-17
BANGLADESH

485
Bangladesh University of Engineering and Technology
Dacca-2
BANGLADESH

486
Canadian Hunger Foundation
ATTN: Project Director
P.O. Box 27
Igbal Road, Chittagong
BANGLADESH

487
Canadian International Development Agency
(CIDA)
1st Secretary, Development
House 69, Road 3, Dhanmondi
Dacca
BANGLADESH

488
Canadian University Service Overseas (CUSO)
37, Indira Road
P.O. Box 404
Dacca
BANGLADESH

489
Chittagong University Rural Development Project
Dr. Moinuddin Ahmed Kahn, Chairman
Dept. of Islamic History & Culture
Chittagong University, Chittagong
BANGLADESH

490
CORR Grain Storage Project
Jalchatra P.O.
Tangail District
BANGLADESH

491
Cottage Industry Project
Islamic University
Santosh, Tangail
BANGLADESH

492
Danish International Development Agency
(DANIDA)
G.P.O. Box No. 349
Dacca
BANGLADESH

493
Gono, Shasthaya Kendra
Nayarhat District
BANGLADESH

494
HEED Bangladesh
240, Road 21, Dhanmondi, Dacca 5
P.O. Box 5052, Newmarket
Dacca 5
BANGLADESH

495
International Voluntary Service (IVS)
549F, Road 14
Dhanmondi
Dacca 5
BANGLADESH

496
Jatio Church Parisad Bangladesh
Emergency Food Programme Relief and Development Corps
P.O. Box 220, Ramna
Dacca 2
BANGLADESH

497
Mennonite Central Committee
Box 785
Dacca-2
BANGLADESH

498
Mennonite Central Committee (MCC)
Director
1-1 Block A, Mohammadpur

Mirpur Road, G.P.O. Box 785
Dacca-2
BANGLADESH

499
National Board of Bangladesh Women's Rehabilitation Program
Chairman
104 Road 3, Dhanmondi
Dacca
BANGLADESH

500
Organisation of Netherlands Volunteers
296 Elephant Road
Dacca
BANGLADESH

501
Peoples Health Centre
via Dhamria
Dacca District
BANGLADESH

502
Ratanpur Agricultural Implement and Training Center
Village Ratanpur
P.O. Sr. Ratanpur
Dist. Kushita
BANGLADESH

503
The Society for Those Who Have Less
Cirant Chalk
Char Fassion Orphanage
Char Fassion
Bhola Island, Barisal Dist.
BANGLADESH

504
Women's Career Training Institute
Director
New Bailey Road
Dacca
BANGLADESH

FIJI

505
Ministry of Commerce, Industry and Cooperatives
Development Bank Centre
Victoria Parada
Suva
FIJI

506
University of the South Pacific
School of Natural Resources
Lavcala Bay
P.O. Box 1168
Suva
FIJI

GILBERT ISLANDS

507
The Tungavala Society
P.O. Box 231
Bikenibeu, Tarawa
GILBERT ISLANDS

HONG KONG

508
Hong Kong Productivity Centre
ATTN: Head, Information Center
G.P.O. Box 6132
HONG KONG

509
Hong Kong Standards and Testing Centre (HK-STC)
Eldex Industrial Building
12/F, Unit A
21 Ma Tau Wei Road
Hung Hom, Kowloon
HONG KONG

INDIA

510
Action for Agricultural Renewal in Maharashtra (AFARM)
3 Victoria Road
Poona 1
Maharashtra
INDIA

511
Action for Food Production (AFPRO)
Community Center
C-17 Safdarjung Development Area
New Delhi-110016
INDIA

512
Action for Food Production
c-52 N.D. South Extension 11
New Delhi 110049
INDIA

513
Action Research Centre for Entrepreneurship
c/o MMC School of Management
Third Floor, Court Chambers
New Marine Lines
Bombay 400 020
INDIA

514
Afro-Asian Rural Reconstruction Org.
c/118 Defence Colony
New Delhi 24
INDIA

515
Agricultural Implements & Power Devel. Centre
Allahabad Agricultural Institute
Allahabad, U.P.
INDIA

516
Agricultural Tools Research Center
Mohan Parikh, Suruchi Campus
P.O. Box 4
Bardoli 394 601
Gujarat
INDIA

517
Ahmedabad Textile Industries Research Association
Polytechnic
Ahmedabad 15
INDIA

518
All India Khadi and Village Industries Commission (KVIC)
Gramodaya
3 Irla Road
Vile Parle (West
Bombay 400 056
INDIA

519
Allied Publishers, Pvt. Ltd.
15 Graham Road, Ballard Est.
Bombay 400 038
INDIA

520
Amarpurkashi Village Project
P.O. Bilari
Dist. Moradabad, U.P.
INDIA

521
Anand Niketan Ashram
P.O. Rangpur via Kosendra
District Baroda, Gujarat
INDIA

522
Application of Science and Technology to Rural Development (ASTRA)
Indian Institute of Science
Bangalore, 560 0012
INDIA

523
Appropriate Technology Cell
Allahabad Polytechnic
Allahabad 211002
Uttar Pradesh
INDIA

524
Appropriate Technology Cell
Ministry of Industrial Development
Government of India
268 Udyong Bhavan
New Delhi
INDIA

525
Appropriate Technology Development Association
C1-/1 River Bank Colony
Lucknow
Uttar Pradesh
INDIA

526
Appropriate Technology Development Association (ATDA)
Post Box 311
Gandhi Bhawan
Lucknow 226 001
Uttar Pradesh
INDIA

527
Appropriate Technology Development Group
B IV/36 Safdarjung Development Scheme
New Delhi 16
INDIA

528
Appropriate Technology Development Unit
Gandhian Institute of Studies
P.O. Box 116

Rajghat, Varanasi 221001, U.P.
INDIA

529
Appropriate Technology Unit
Dept. of Industrial Development
Ministry of Industry
Udyog Bhavan
New Delhi 110011
INDIA

530
Appropriate Technology Unit
Indian Institute of Technology
Powai
Bombay 400076
INDIA

531
Ap-Tech Newsletter
P.O. Box 311
Gandhi Bhawan
Mahatma Gandi Road
Lucknow-226 001
INDIA

532
Association of Indian Engineering Industry
6 Netaji Subhas Rd.
Calcutta, 700 001
INDIA

533
Association of Scientific Workers of India
10 Rajendra Park
New Delhi 110 060
INDIA

534
Association of Voluntary Agencies for Rural Development (AVARD)
A 1, Kailash Colony
New Delhi 110 048
INDIA

535
ASTRA-Cell
Department of Inorganic and Physical Chemistry
Indian Institute of Science
Bangalore 56 00 12
INDIA

536
Auma Research & Development Facility
Auroville Centre for Environmental Studies
3 rue Dupuy, Pondicherry
605002
INDIA

537
Bafe Lab (Pvt) Ltd.
795/54, Datar Building
Deccan Gymkhana
Poona 411 004
INDIA

538
Caritas India
C.B.C.I. Centre
Ashok Place
New Delhi 110 001
INDIA

539
Central Arid Zone Research Institute
Jodhpur
Rajasthan
INDIA

540
Central Building Research Institute
Roorkee U.P.
INDIA

541
Central Food Technological Research Institute
Cheluramba Mansion
V.V. Mohalla P.O.
Mysore 570013
INDIA

542
Central Glass and Ceramic Research Institute
Jadavpur
Calcutta 32
INDIA

543
Central Indian Medicinal Plant Organisation
Haldwani, U.P.
INDIA

544
Central Leather Research Institute
Adyar
Madras 600 020
INDIA

545
Central Mechanical Engineering Research Institute
Mahatma Gandhi Avenue
Durgapur 713209
West Bengal
INDIA

546
Central Road Research Institute
Delhi-Mathura Rd.
P.O. C.R.R.I.
New Delhi 110020
INDIA

547
Central Soil Salinity Research Institute
ATTN: R.C. Mondal
Karnal, Haryana
INDIA

548
Centre for Development Studies
Aakulam Road
Ulloor
Trivandrum 695 011
Kerala
INDIA

549
Centre for Science in the Village
Magian Sangrahalaya
Wardha 442001
Maharashra
INDIA

550
"Compost Vingnanam"
Tenkasi 627 811
29 New Street
Tirunwelveli Dist.
Tamil Nadu
INDIA

551
Constellate Consultants, Ltd.
5 Anand Lok
New Delhi 110 049
INDIA

552
Council of Scientific and Industrial Research
New Delhi 110001
INDIA

553
Deena Bandu Rural Centre
R.K. Pet 631303
Tamil Nadu
INDIA

554
Development Commissioner for Small-Scale Industry
Nirman Bhavan
New Delhi 110011
INDIA

555
Directorate General of Technical Development
UDYOG Bhawan
New Delhi
INDIA

556
District Industries Centre
Haridas Chatterjee Luke
Gaya 823001, Bihar
INDIA

557
Dryland Agricultural Institute
P.O. Kanke
Ranchi, Bihar
INDIA

558
Forest Products Research Institute
Dahradum
Uttar Pradesh
INDIA

559
Forum of Industrial Technologists (F.I.T.)
1233 A, Apte Road
Poona 411 004
Mahrashtra
INDIA

560
Foundation for Research in Community Health
Dhekawade
P.O. Awas
Alibag Taluk
Dist. Kolaba
Maharashtra
INDIA

561
Friends' Rural Development Centre
Rasulia
Hoshangabad, M.P.
INDIA

562
Gandhian Institute of Studies
Appropriate Technology Development Unit
Post Box No. 116
Rajghat, Varanasi-221001
INDIA

563
Garlick Engineering
Bapuras Jagtap Marg.
Sant Gade Maharaj
Chowk, Bombay
INDIA

564
Gobar Gas Experimental Station
Ajitmal
Etawah U.P.
INDIA

565
Gobar Gas Research and Development Centre
Khadi and Village Industries Commission
Kora Gramodyog Kendra
Borivli
Bombay 92
INDIA

566
Government College of Engineering
Salem College
Tanulnada
INDIA

567
Gram Seva Samti
Taronda Mitaya
P.O. Raisalpur
District Hoshangabad
Madhya Pradesh
INDIA

568
Higginbothams, Ltd.
165 Anna Salai
Madras 600 002
INDIA

569
Hydroponic Advisory & Information Unit
P.O. Box 31
Bombay 1
INDIA

570
Home Science College
SRI Anivanshilingam
Colmbatore 11
INDIA

571
Indian Agricultural Research Institute
Pusa
New Delhi
INDIA

572
Indian Banks Association
Stadium House
Block 3, 6th Floor
Veer Nariman Road
Bombay 400 020
INDIA

573
Indian Grain Storage Institute
Grain Storage Research Centre
Hapur
Uttar Pradesh
INDIA

574
Indian Institute of Petroleum
P.O. O.E.P.
Dhera Dun
248005, U.P.
INDIA

575
Indian Institute of Science
Cell for the Application of Science & Technology to Rural Areas
Dept. of Inorganic and Physical Chemistry
Bangalore 560012
INDIA

576
Indian Institute of Science
Dept. of Chemical Engineering
Bangalore 12
INDIA

577
Indian Institute of Science
Department of Mechanical Engineering
Bangalore 560012
INDIA

578
Indian Institute of Sugar Cane Research
Lucknow 226002
INDIA

579
Indian Institute of Technology
Dept. of Metallurgical Engineering
Kanpur 208016
INDIA

580
Indian Institute of Technology/Madras
Engineering Research Center
Madras
INDIA

581
Indian Institute of Technology/New Delhi
Mechanical Engineering Department
New Delhi-110029
INDIA

582
Indian Institute of Technology
Kharagpur
West Bengal
INDIA

583
Indian National Buildings Organisation
Nirman Bhavan
New Delhi 1
INDIA

584
Indian Plywood Industries Research Institute
Bangalore
INDIA

585
Indian Refrigeration Industries, Ltd.
Singanalore
Coimbatore 5
Tamil Nadu
INDIA

586
Institute of Engineers
8, Gokhale Road
Calcutta 20
INDIA

587
Institute of Social Service
Nirmala Niketan
38 New Marine Lines
Bombay 400 020
INDIA

588
Institute of Technology
Banaras Hindu University
Varanasi 121005
INDIA

589
International Crops Research Institute for the Semi-Arid Tropics (ICRISAT)
1-11-256, Begumpet
Hyderabad 500 016
Andra Pradesh
INDIA

590
Jamnalal Bajaj Central Research Institute for Village Industries
Maganwadi
Post Box 4
Wardha, Maharashtra
INDIA

591
Jute Technology Research Institute
Calcutta
INDIA

592
Kamla Nehru Institute of Science and Technology
Sultanpur U.P.
INDIA

593
Khadi & Village Industries Commission
"Gramodaya"
3 Irla Road
Vile Parle (West)
Bombay 400 056 A.S.
INDIA

594
Khadi & Village Industries Commission, "Gramodaya"
ATTN: The Director
Directorate of Gobar Gas Scheme
Irla Road, Vile Parle (West)
Bombay 400 056
INDIA

595
Kishore Bharati

Post Malhanvada via Bankhedi
P.O. Box 25
District Hoshangabad
M.P. 461 990
INDIA

596
Larsen and Toubre Ltd.
Eros Bldg. 4th Fl.
Churchgate
Bombay 400 020
INDIA

597
The Learners Center
Casa Leao
Cobrovaddo
Calangute
P.O. Goa
INDIA

598
Madurai Windmill Committee
69 P.T. Rajan Road
Madurai
Tamil Nadu 625 002
INDIA

599
Mandya National Paper Mills
Belagula-Karnataka
INDIA

600
Mechanical Engineering Research
and Development Organisation
(MERAD)
Madras
INDIA

601
Medico Friend Circle
Rajghat
Varanasi 221 001
INDIA

602
Ministry of Industry and Supply
Small Industries Service Institute
Government of India
Karan Nagar
Srinagar 190 010
INDIA

603
Mitraniketan
Vellanadu
Nedumangad, District Trivandrum
Kerala 695 543
INDIA

604
Murugappa Chettiar Research Centre
Thara Mani
Madras 600 042
INDIA

605
Mysore Acetate and Chemicals
Mandya-Karnataka
INDIA

606
National Environmental Engineering
Research Inst.
Nehru Marg
Nagpur 440 020
INDIA

607
National Industrial Development Corp.
Ltd.
P.O. Box 5212
New Delhi 110 021
INDIA

608
National Research Development Corp.
of India
61 Ring Road, Lajpat Nagar III
New Delhi 110 024
INDIA

609
National Sugar Institute
P.O.: N.S.L.
Kanpur 17
INDIA

610
NOVA 2000, Consultants and Designers
D16 Kailash Colony
New Delhi 110 048
INDIA

611
Oil Technological Research Institute
Anantapur
Pradesh
INDIA

612
Peace Corps
5 Mahatma Gandhi Marg
Kilokri

New Delhi
INDIA

613
Planning Research & Action Division
State Planning Institute
Kalakankar House
Lucknow
INDIA

614
Protein Foods & Nutrition Development Assn. of India
22 Bhulabhai Desai Road
Bombay 400 026
INDIA

615
Punjab Agricultural University
Dept. of Mechanical Engineering
College of Agricultural Engineering
Ludhiana, Punjab
INDIA

616
A.S. Venkat Rao, F.R.H.S.
28 New Street
Tenkasi 627811
Tamil Nadu
S. INDIA

617
Regional Research Laboratory
Jorhat-785 006
Assam
INDIA

618
Regional Research Laboratory
Canal Road, Jammu-Tawi
Jammu
INDIA

619
Resources Development Institute
Bhopal, Madhya Pradesh
INDIA

620
Saghan Kshetra Samiti
Sewapuri
Varanasi
Uttar Pradesh
INDIA

621
Sara Technical Services, Pvt. Ltd.
9 Vasant Vihar
Commercial Centre 11
New Delhi 110 057
INDIA

622
Saran Engineering Co. Ltd.
Sutherland House
Kanpur, U.P.
INDIA

623
Sethi
National Industrial Development Corp. Ltd.
Chanakya Bhavan
Vinay Marg
New Delhi 110 021
INDIA

624
Shahdol Appropriate Technology Unit
Vidushak Karkhana
Anuppur
District Shahdol
MP 484 224
INDIA

625
Shri Ram Institute for Industrial Research (SRIFIR)
19 University Road
Delhi 110 007
INDIA

626
Skills for Progress (SKIP)
72, Brigade Road
Bangalore 560 025
INDIA

627
Small Industry Extension Training Institute (SIET)
Yousufguda
Hyderabad, 500 045
A.P.
INDIA

628
Small Scale Industries
Office of Development Commission
Nirman Bhavan (South Wing) 7th Floor
Maulana Azad Marg
New Delhi 110 011
INDIA

629
Solar Energy

Technology Bhavan
New Delhi 110 029
INDIA

630
Structural Engineering Research Centre
Roorke
INDIA

631
Sugar Technologists Association of India
Kalianpur, Kanpur 17
INDIA

632
Tadaypur University
Dept. of Food Technology & Biochemical Engineering
Calcutta 32
INDIA

633
Tata Energy Research Institute
Bombay House 24 Homi Mody St.
Bombay 400 023
INDIA

634
Technology Bhavan
Dept. of Science & Technology
New Mehrauli Road
New Delhi 110 029
INDIA

635
Terre des Hommes
Adi D Patel
19 Sahney Sujan Park
Kondwa Rd.
Pune 411 001
INDIA

636
Tibetan Mission House
Kalimpong
Darjeeling District
West Bengal
INDIA

637
Utkal Navajivan Mandal
Angul P.O. District Dhenkanal
Orissa 759 122
INDIA

638
Vaikunthbhai Mehta Smarak Trust
5th Floor NKM International House
178 Backbay Reclamation
Churchgate
Bombay 400 020
INDIA

639
Venkat Rao
28, New Street
Tenkasi (627, 811)
Tirunelveli District
Tamil Nadu
S. INDIA

640
Vigyan Shiksha Kendra
(Science Education Centre)
Atarra, Banda 210 201
INDIA

641
Village Industrialisation Association
"Pragati"
Civil Lines
Wardha, M.S.
INDIA

642
Village Reconstruction Organisation
ATTN: Father/Prof. Michael Windey
6/9 Brodipet
Guntur
Andra Pradesh 522 002
INDIA

643
Vimala Welfare Centre
Vimalalayam
Eruakulam
Cochnin 18
Kerala State
INDIA

644
Walchandnagar Industries Ltd.
Walchandnagar, Poona
Maharashtra
INDIA

645
Water Development Section
P.O. Badal Mission
Tal. Newasa District
Maharashtra
INDIA

646
Water Development Society

C-2 and C-5 Industrial Estate
Moula Ali
Hyderabad 500 040 A.P.
INDIA

647
Xavier Institute
Food Marketing Centre
P.O. Box 47
Jamshedpur-1
INDIA

648
Yantra Vidyalaya Agro-Industrial Service Center
Faculty of Gandhi Vidhapith, Vedchhi
Suruchi Campus, P.O. Box 4
Bardoli-394 601
Surat Dist.
INDIA

649
Youth's Action (India)
Central Office
19/22 Shakti Nagar
Delhi 110 007
INDIA

650
Youth's Action (India)
Rural Programme
P.O. Sadarpur Dist.
Sultanpur, U.P.
INDIA

INDONESIA

651
Badan Urusan Tenaga Keya Sukatela Indonesia (BUTSI)
Jalan Halimun 4
Jakarta
INDONESIA

652
Balai Penelitian Perkebunan Borgor (BPPB)
Research Institute for Estate Crops
P.O. Box 05
Bogor
INDONESIA

653
Bandung Institute of Architecture
Strategy for Planning, Design & Development Study Group
Dept. of Architecture
Bandung
INDONESIA

654
Building Research Institute
Directorate General of Housing, Building, Planning, & Urban Development
Ministry of Public Works and Electric Power
Jalan Tamansari 124
P.O. Box 15
Bandung
INDONESIA

655
The Central Research Institute for Agriculture (LP3)
Jl. Merdeka 99
Bogor
INDONESIA

656
Development Technology Centre (DTC)
Institute of Technology Bandung (ITB)
Jalan Ganesha 10
P.O. Box 276
Bandung
INDONESIA

657
Indonesian Institute of Sciences (LIPI)
Jl. Teuku Chik
Ditiro 43, Jakarta
P.O. Box 250
Jakarta
INDONESIA

658
Indonesian Institute of Sciences
Jln. Teuku Cik Ditiro 43
P.O. Box 250
Jakarta
INDONESIA

659
Institute for Economic and Social Research
Faculty of Economics
University of Indonesia
4 Salemba
Djakarta
INDONESIA

660
Institute Pertanian Bogor
Dept. of Agricultural Engineering
Bogor Agricultural University

Fatemeta Jalan Gunung Gede
TLP BOT 571
Bogor
INDONESIA

661
Institute of Rural and Regional Studies
University of Gadjahmada
Bulaksumur-E-12
Yogyakarta
INDONESIA

662
I.P.B.
Department of Agriculture Productivity and Technology
Fatemeta-Jalan Gunung Gede
Bogor
INDONESIA

663
Lembaga Ilmu Pengetahaun Indonesia (LIPI)
Djl Teuku Tjhik Ditiro 43
Jakarta
INDONESIA

664
Lembaga Penelitian Dan Pendidikan Industri (LPPI)
Jl. Tulondong Bawah 11/14, Blok. R
Kabayoran Baru
P.O. Box 2802
Jakarta
INDONESIA

665
Proyek Teknologi Tepat (PTT)
Dian Desa
Jalan Kerto Muja Muju 8
Yogyakarta
INDONESIA

666
Pusat Documentasi Ilmiah Nasional (PDIN)
Jl. Jenderal Gatot Subroto
P.O. Box 3065/Jkt.
Jakarta
INDONESIA

667
Regional Adaptive Technology Center
Hasanuddin University RATC
Kampus Baraya
Ujungpandang
INDONESIA

668
The Research Institute for Industrial Crops (LIPI)
J.L. Cimanggn 1
Bogor
INDONESIA

669
Anton Soedjarwo
Jalan Pacar A 67
Jogjakarta
INDONESIA

670
Strapp-Group
Department of Architecture
I.T.B.
Jalan, Ganosha 10, Bandung
INDONESIA

671
The Sugarcane Research Institute (BP3G)
J.L. Pahlawan 25
Pasuruan
INDONESIA

672
Tanjungsari Academy Agriculture
Industrial Development Division
APT SPMA Tanjungsari
Kabupaten Sumegand
INDONESIA

673
Village Technology Unit Butsi
Jalan Halimun 4
Jakarta
INDONESIA

674
Voluntary Agencies for Rural Development
Badan Urusan Tenaga Keya Sukatela Indonesia
Jalan Halimun 4
Jakarta
INDONESIA

IRAN

675
Centre for Endogenous Development Studies
Alashtar, lorestan
IRAN

676
Development Workshop
224 Saba Shomali
Tehran
IRAN

677
Ecodevelopment Cluster
Bu-Ali Sina University
Hamadan
IRAN

678
Institute of Standards and Industrial Research of Iran (ISIRI)
Karadj, IRAN
P.O. Box 2937
Tehran
IRAN

679
Ministry of Industries and Mines
Organisation for Small-Scale Industries and Industrial Estates of Iran
Tehran
IRAN

680
Small-Scale Industries Organisation of Iran
P.O. Box 3285
Tehran
IRAN

JAPAN

681
Asian Productivity Organisation (APO)
Aoyania Daiichi Mansions 4-14
Akasaka 8-Chome
Minato-ku
Tokyo 107
JAPAN

682
CECOCO
Chuo Boeki Goshi Kaisha
P.O. Box 8
Ibaraki City
Osakea Pref
567 JAPAN

683
Federal Institute of Industrial Research (FIIR)
FIIR, Oshodi
FIIR, P.M.B. 1023
Ikeja Airport
JAPAN

684
Japan Industrial Technology Association
20-Mort Bldg. 8F
2-7-4-Nishi Shinbashi
Minato-ku
Tokyo 105
JAPAN

685
Japan Institute of Oils & Fats
Other Foods Inspection Foundation
2708 Nikon Gashi
Hama-cho 3-Chome
Chuo-ku, Tokyo
JAPAN

686
OISCA International
612 Izumi 3-chome
Suginami-Ru
Tokyo 168
JAPAN

687
UN University
Toho Seimei Bldg, 15-1
Shibuya 2 Chome
Shibuya-ku
Tokyo
JAPAN

MALAYSIA

688
Bokit Lan Methodist Rural Development Programme
P.O. Box 155
Sibu, Sarawak
MALAYSIA

689
Institute Pertanian Bogor
Fatemeta Jalan Gunung Gede
TLP BOT 571
Bogor
MALAYSIA

690
Malaysian Agricultural Research & Development Institute (MARDI)
Jalan Marktab
Kuala Lumpur
MALAYSIA

691
Malaysian Agricultural Research & Development Institute
P.O. Box 208
Sungai Besi
Serdang
Selangor
MALAYSIA

692
National Institute for Scientific and Industrial Research (NISIR)
Shah Alam
Selangor
MALAYSIA

693
National Productivity Centre
P.O. Box 64-Sultan St.
Petaling Jaya, Selangor
MALAYSIA

694
National Scientific Research & Development Council
Ministry of Science, Technology and Development
Tingket-14
Bangnnan, Oriental Plaza
Jalandarry, Kula Lumpur 0401
MALAYSIA

695
Standards & Industrial Research Institute of Malaysia (SIRIM)
Lot 10810, Phase 3
Federal Highway
Shah Alam
Selangor
MALAYSIA

696
University Pertanian Malaysia
P.O. Box 203
Sungai Besi
Selangor
W. MALAYSIA

MONGOLIA

697
Faculty of Civil Engineering
Mongolian State University
MONGOLIA

NEPAL

698
Agricultural Projects Services Center
Dillibazar
Kathmandu
NEPAL

699
Balaju Yantra Shala (P) Ltd. (BYS)
P.O. Box 209
Balaju
Kathmandu
NEPAL

700
British Gurkha Ex-Servicemen Reintegration Training Scheme
Lumle Agricultural Centre
c/o P.O. Box 1
Pokhara
Gandaki Anchal
NEPAL

701
Butwal Technical Institute
Butwal
via P.O. Nuatanwa
NEPAL

702
Canadian International Development Agency
c/o Institute of Medicine
Box 1240
Kathmandu
NEPAL

703
Industrial Services Centre
P.O. Box 1318
Kathmandu
NEPAL

704
Institute of Applied Science and Technology
Central Campus
Tribjuvan University
P.O. Box 4
Dharan
NEPAL

705
Research Centre for Applied Science and Technology (RECAST)
Tribhuvan University
Kirtapur, Kathmandu
NEPAL

706
Swiss Association for Technical Assistance (SATA)
P.O. Box 113
Kathmandu
NEPAL

707
UNICEF
P.O. Box 1187
Lainchur
Kathmandu
NEPAL

708
United Mission to Nepal
P.O. Box 126
Kathmandu
NEPAL

NEW CALEDONIA

709
South Pacific Commission (SPC)
Post Box D5
Noumea Cadex
NEW CALEDONIA

NEW HEBRIDES

710
Kristian Institute of Technology (KITOW)
P.O. Box 16
Isaugel
Tanna via Villa
NEW HEBRIDES

NEW ZEALAND

711
Department of Scientific and Industrial Research
Physics & Engineering Laboratory
Private Bag
Lower Hutt
NEW ZEALAND

712
Lincoln College Energy Research Group
Canterbury
NEW ZEALAND

713
Productivity Centre
Department of Trade and Industry
Private Bag
Wellington
NEW ZEALAND

PAKISTAN

714
Agricultural Research Institute
Tandojam
Sind
PAKISTAN

715
Appropriate Technology Centre
506 P Block
President's Secretarian Planning Division
Government of Pakistan
Islamabad
PAKISTAN

716
Appropriate Technology Development Organization
Government of Pakistan
1-B St. 47 F-7/1 PO Box 1306
Islamabad
PAKISTAN

717
Church of Pakistan
113 Quasim Road
P.O. Box 204
Multan Cantt
PAKISTAN

718
Industrial Development Bank of Pakistan
Planning, Evaluation & Research Department Library
Post Box No. 7258
Karachi 3
PAKISTAN

719
Irrigation Research Institute
Lahore
PAKISTAN

720
Irri-Pak Agricultural Machinery Program
Agricultural Machinery Programme
73-A Satellite Town
Rawalpindi
PAKISTAN

721
Ministry of Food, Agriculture and Underdeveloped Areas
Food and Agriculture Division
Planning Unit
Islamabad
PAKISTAN

722
Pakistan Academy for Rural Development
Peshawar
North West Frontier Province
PAKISTAN

723
Pakistan Council of Scientific and Industrial Research
PCSIR
21 E Block 6
P.E.C.H.S.
Karachi 29
PAKISTAN

724
Pakistan Scientific and Technological Information Center
PASTIC
13-P Almarkaz, F-7/2
P.O. Box 1217
Islamabad
PAKISTAN

725
Pakistan Small Industries Corporation
Lahore
PAKISTAN

726
Resource Development Corp.
8th Floor, Dawood Center
P.O. Box 3724
Karachi 4
PAKISTAN

PAPUA NEW GUINEA

727
Association of Technologies of P.N.G.
Box 1358
Boroko
PAPUA NEW GUINEA

728
Committee on Rural Development
University of Technology
P.O. Box 793
Lae
PAPUA NEW GUINEA

729
Department of Agriculture Stock and Fisheries
P.O. Box 2417
Konedobu
PAPUA NEW GUINEA

730
Department of Business Devel.
Small Industries Division
Post Office Wards Strip
Waigani
PAPUA NEW GUINEA

731
Department of Labour and Industry
Industrial Projects
P.O. Box 5644
Boroko
PAPUA NEW GUINEA

732
Industrial Development Education
P.O. Alekishafen
Madang District
PAPUA NEW GUINEA

733
Liklik Buk Information Centre
Box 1920
Lae
PAPUA NEW GUINEA

734
Melanesian Council of Churches
P.O. Box 1920
Lae
PAPUA NEW GUINEA

735
Office of Village Development of the Dept. of the Prime Minister
Boroko
PAPUA NEW GUINEA

736
Purari Action Group
c/o National Parks
P.O. Box 2749
Boroko
PAPUA NEW GUINEA

737
Reading Methods & Materials Centre

University of Papua New Guinea
P.O. Box 4820
Post Moresby
PAPUA NEW GUINEA

738
South Pacific Appropriate Technology Foundation
P.O. Box 6937
Boroko
PAPUA NEW GUINEA

739
Technology Development Unit
University of Technology
P.O. Box 793
Lae
PAPUA NEW GUINEA

740
Village Technology Information Centre
Department of Environment
P.O. Wards Strip
Port Moresby
PAPUA NEW GUINEA

741
Yanpela Didiman Association
P.O. Box 39
Banz
Western Highland Province
PAPUA NEW GUINEA

PHILIPPINES

742
Asian Development Bank
P.O. Box 789
Manila
PHILIPPINES

743
Asian Regional Training & Development Organization
Ground Floor, 1551 Building
University of the Philippines Campus
Diliman Quezon City 3004
PHILIPPINES

744
CENDHRRA-Center for the Development of Human Resources in Rural Asia
P.O. Box 458 Greenhills
San Juan, Metro Manila
PHILIPPINES

745
Centre for Research and Communication
Manila
PHILIPPINES

746
Council for Asian Manpower Studies, Ltd.
P.O. Box 6, U.P. Post Office
Diliman, Quezon City 3004
PHILIPPINES

747
Department of Industry
Chronicle Bldg., 3rd Floor
Meralco Ave.
Pasig, Rizal
PHILIPPINES

748
Development Academy of the Philippines
3rd Floor, B.F. Condominium Bldg.
Aduana Street
Intramuros
Manila
PHILIPPINES 2801

749
Economic Development Foundation, Inc.
P.O. Box 1896
Manila
PHILIPPINES

750
Federation of Free Workers
Room 303, Cruz Building
1452 Taft AVenue
Manila 1208
PHILIPPINES

751
Institute for Small-Scale Industries
East Jacinto St., U.P. Campus
Diliman, Quezon City D-505
PHILIPPINES

752
International Institute for Rural Reconstruction (IIRR)
Silang Cavite
PHILIPPINES 2720

753
International Rice Research Institute (IRRI)

P.O. Box 933
Manila
PHILIPPINES

754
Metals Industry Research and Development Center (MIRDC)
5th Floor, Ortigas Building
Ortigas Ave.
Pasig, Rizal
PHILIPPINES

755
National Economic Development Authority
Industrial Programs Office
P.O. Box 1116
Manila
PHILIPPINES

756
Regional Adaptive Technology Center
Mindanao State University
Marawi City, Mindanao
PHILIPPINES

757
Scientific Library and Documentation Division
National Science Development Board
P.O. Box 3596
Manila
PHILIPPINES 2801

758
Tech Tran Corporation
7th Floor, Merchants Realty Bldg.
313 Buendia Ave., Makati
Metro Manila
PHILIPPINES

759
Technology Resources Center (TRC)
TRC Building
Buendia Avenue Extension
Makati, Metro Manila
PHILIPPINES

760
University of the Philippines
ISSI-Institute for Small-Scale Industries
Virata Hall, Diliman
Quezon City
PHILIPPINES

REPUBLIC OF CHINA (TAIWAN)

761
Bureau of Public Works
Department of Reconstruction
Taiwan Provincial Government
4 Kai Feng Street, Sec. I
Taipei
TAIWAN

762
China Agricultural Machinery Co., Ltd.
11 Tung Hsing St.
Taipei 105
TAIWAN

763
The Chinese Institute of Engineers
30-1 Ai Kou W. Road
TAIWAN

764
Joint Commission on Rural Reconstruction
Animal Industry Division
37 Nan Hai Road
Taipei, 107
TAIWAN

765
The Land Bank of Taiwan
46 Kuan Chien Road
Taipei
TAIWAN

766
Science and Technology Information Center
National Science Council
P.O. Box 4, Nankang
Taipei
TAIWAN

767
Taipei Architect Association
2nd Floor, 3-1 Lane 3
Chang An W. Road
Taipei
TAIWAN

768
Taiwan Sugar Corporation
Room 621
25 Pao Ching Road
Taipei
TAIWAN

769
Union Industrial Research Institute

(UIRI)
1021 Kuang Fu Road, Hsinchu
P.O. Box 100, Hsinchu
TAIWAN

770
United Association for Contractors
2nd Floor, 2 Lane 84
Yung Ping S. Road
Taipei
TAIWAN

SINGAPORE

771
Singapore Institute of Standards & Industrial Research (SISIR)
179 River Valley Road
P.O. Box 2611
SINGAPORE

772
Techonet Asia-Asian Network for Industrial Technology Information and Extension
Tanglin P.O. Box 160
SINGAPORE

SOUTH KOREA

773
Federation of Korean Industries
28th Floor, Samilro Bldg.
10, Kwanchol-Dong, Changro-Ku
Seoul
KOREA

774
Korea Institute of Science and Technology (KIST)
39-1, Hawolgok-dong
Sungbuk-ku, Seoul
P.O. Box 131
Cheong Ryang, Seoul
SOUTH KOREA

775
Korea Institute of Science and Technology
Technology Transfer Centre
P.O. Box 131
Cheong Ryang
Seoul
SOUTH KOREA

776
Korea Scientific & Technological Information Center (KORSTIC)
ATTN: Director Dept. of Information Resources
C.P.O. Box 1229
Seoul
SOUTH KOREA

777
Korea-U.K. Farm Machinery Training Project
Office of Rural Development
Suweon
SOUTH KOREA

778
Medium Industry Bank
Extension Services
36-I, 2KA, Ulchiro
Seoul
SOUTH KOREA

779
Soong Jun University
Integrated Development Center
ATTN: Director
Sang Do Dong
Seoul 151
SOUTH KOREA

780
Yeungnam University
Regional Adaptive Technology Center (RATC-UY)
Kyungsan 632
SOUTH KOREA

SRI LANKA

781
Agricultural Research and Training Institute (ARTI)
Elibank Road
Colombo 8
SRI LANKA

782
Appropriate Technology Group
c/o Chemical Industries Ltd.
P.O. Box 352
Colombo 1
SRI LANKA

783
Ceylon Institute of Scientific and Industrial Research
P.O. Box 787
Colombo 7
SRI LANKA

784
Cooperative Wholesale Establishment
Research Development Division
21 Vauxale Street
Colombo 2
SRI LANKA

785
Department of Highways
Research Laboratory
Ratmalana
SRI LANKA

786
Industrial Development Board
615 Galle Road
Katubadde, Moratuwa
SRI LANKA

787
Lanka Jatika Sarvodaya Shramadana Sangamaya Inc.
77 De Soysa Road
Moratuwa
SRI LANKA

788
Marga Institute
P.O. Box 601
61, Isipathana Mawatha
Colombo
SRI LANKA

789
Ministry of Planning and Economic Affairs
Regional Development Division
P.O. Box 1532
Colombo
SRI LANKA

790
National Engineering Research & Development Centre of Sri Lanka (NERD Centre)
285 Galle Road
Colombo 3
SRI LANKA

791
National Engineering Research & Development Centre
Katubedde Campus
University of Sri Lanka
Katubedde
SRI LANKA

792
Rubber Research Institute of Ceylon
Dartonfield Group
Agalawatta
SRI LANKA

793
Sarvodaya Appropriate Technology Development Programme
Lanka Jatika Sarvodaya Schramadana Sangamaya (Inc.)
77 de Soysa Road
Moratuwa
SRI LANKA

794
Sarvodaya Shramadana Sangamaya
Tanamalwila
SRI LANKA

795
Serendeepam Community Development Project
Sithankerny N.P.
SRI LANKA

796
Small Industries Development
Hemas Building
Colombo 1
SRI LANKA

797
Small Scale Hydro Plant
Dept. of Civil Engineering
Katubedde Campus
University of Sri Lanka
Katubedde
SRI LANKA

798
Solar/Wind Energy
Electricity Board
Colombo
SRI LANKA

799
Sri Lanka Association for the Advancement of Science (SLAAS)
c/o Regional Development Div.
Ministry of Planning and Economic Affairs
P.O. Box 1532
Colombo
SRI LANKA

800
The Sri Lanka Center for Development Studies
17 Park Avenue

Colombo 5
SRI LANKA

801
Sri Lanka Scientific and Technical Information Centre
National Science Council of Sri Lanka
47/5 Maintland Place
Colombo 7
SRI LANKA

802
University of Sri Lanka
Department of Civil Engineering
Katubedda Campus
Moratuwz
SRI LANKA

803
University of Sri Lanka
Department of Mechanical Engineering
Peradeniya Campus
SRI LANKA

THAILAND

804
Applied Scientific Research Corporation of Thailand
196, Phaholyothin Road
Bang Khen
Bangkok 9
THAILAND

805
Asian Institute for Economic Development and Planning
Sri Ayohya Road
P.O. Box 2-136
Bangkok
THAILAND

806
Asian Institute of Technology
P.O. Box 2754
Bangkok
THAILAND

807
Association of Siamese Architects
1155 Phaholyothing Road (near Rajakroo Lace)
Bangkok
THAILAND

808
Economic & Social Commission for Africa & the Pacific (ESCAP)
Sala Santitham
Rajdammern Avenue
Bangkok
THAILAND

809
Engineering Institute of Thailand
Engineering Building
Chulalongkorn University
Bangkok
THAILAND

810
Government Housing Welfare Bank
Mansion No. 9
Rajdamnden Avenue
Bangkok
THAILAND

811
Industrial Finance Corp. of Thailand
101 Naret Road
Bangkok 5
THAILAND

812
Industrial Service Institute
Northern Branch
P.O. Box 82
Chiang Mai
THAILAND

813
Ministry of Interior
Housing Bureau
Department of Public Welfare
Asdang Street
Bangkok
THAILAND

814
National Housing Authority
Klong Chan
Bangkok
THAILAND

815
National Institute of Development Administration
Huamark, Bangkapi
Bangkok 10
THAILAND

816
Southeast Asia Technology Company, Limited (SEATEC)
Nai Lert Building

87 Sukhumuit Road
Bangkok
THAILAND

817
Thailand Management Association (TMA)
308 Silom Road
Bangkok
THAILAND

818
UN Economic Commission for Asia & the Far East (ECAST)
Sala Sanitam
Rajadamnern Avenue
Bangkok
THAILAND

819
United Nations Economic Commission for Asia and the Pacific
Division of Industry and Housing
Sala Santitham
Bangkok 2
THAILAND

820
UN Economic and Social Commission for Asia and the Pacific (ESCAP)
Energy Resources Section
Sala Santitham
Bangkok 2
THAILAND

EUROPE

AUSTRIA

821
Institute for Educational and Development Research
Schottenbastei 6
A-1010 Vienna
AUSTRIA

822
Institute for Environmental Research
Elisavethstrasse 11
A8010 Graz
AUSTRIA

823
Transnational Research Centre
Schloss Eichbuchl
A-2801 Katzeldorf
AUSTRIA

824
United Nations Industrial Development Organisation (UNIDO)
Lerchenfelder Strasse 1
A-1070 Vienna
AUSTRIA

BELGIUM

825
Aangepaste Techniek voor Ontwikkelingslanden (ATOL)
Blyde Inkomststraat 9
3000 Leuven
BELGIUM

826
ABR Engineering
Rue du Trone
4-1050 Bruxelles
BELGIUM

827
Association Internationale de Developpement Rural (AIDR)
18-22 Rue du Commerce
Bruxelles, 4
BELGIUM

828
C.E.R.I.A.
1, Avenue Emile-Gryson
Anderlecht, Brussels 7
BELGIUM

829
European Environmental Bureau
31, Rue Vautier
1040 Brussels
BELGIUM

830
Institute for Promotion of Scientific Research in Industry and Agriculture (IRSIA)
Rue de Crayer 6
B-1050 Brussels
BELGIUM

831
International Association for Rural Development
Rue de Commerce 20, Bte 9
B-1040 Brussels
BELGIUM

832
Internationale Bouworde
Naamsestecnweg 573
3030 Herverlee
BELGIUM

833
International Federation of Small and Medium-Sized Commercial Enterprises (FIPMC)

9 Rue Joseph II
1040 Brussels
BELGIUM

834
University of Louvain Mechanics Department
Celestijnenlaan 300A
B 3030 Heverlee
BELGIUM

835
Vereniging Ecologische Land en Tuinbouw
Lage weg 20
B-2241 Halle—Kempen
BELGIUM

DENMARK

836
Akademiet F. de Tekniske
Videnskaber
Launtoftevej 266
2800 Lyngby
DENMARK

837
Bioteknisk Institut
Holbergsvej
10 Kolding 6000
DENMARK

838
Brandstrup Maskinverksted
Vindumvej 174, Brandstrup
8840—Rødekaersbro
DENMARK

839
Byggeteknisk Hojskole
Drejervei 17 1957 KBH V
Karlslunde
DENMARK

840
Christoph Ullmans Maskinfabrik
Herlev Hovedgade 213
2730—Herlev
DENMARK

841
Community Action in Europe
Radsmandsstrade 10A
DK 1407 Copenhagen
DENMARK

842
Danish Building Research Institute
Housing Dept.
SBI Postboks 119
DK-2970 Hørsholm,
DENMARK

843
DTO-Dansk Teknisk Oplysningstjeneste
Information for Industry
Ørnevej 30
2400 Copenhagen Nv.
DENMARK

844
Foreningen Folkets Okoteket
Biblioteksgruppen
Dronningensgade 14
Dk. 1420
Kobenhaven K
DENMARK

845
F.H. Hvelplund
Brund 7700
Thisted
DENMARK

846
Institute F. Samfundsfag
Rosenborggd 15
1130 KBH
DENMARK

847
Institute for Development Research
104 Vesten Voldgade
DK-1552
Copenhagen V
DENMARK

848
International Solar Powerco Ltd.
Rosekaeret 22b
2860 Soborg
DENMARK

849
Laboratorist for Energiteknik
Danmarks Tekniske Højskole
Bygning 403
DK 2800
Lyngby
DENMARK

850
Niels Bohrs Institute

Blegdamsvej 17
2100 KBH Ø
DENMARK

851
NOAH
Radhusstraede 13
1466 Copenhagen K
DENMARK

852
Nyborgvej
5863-Fjerritslen
Fyn
DENMARK

853
ØKO-RA
Rebslagervej 11
2400 Copenhagen NV
DENMARK

854
OOA
Skindergade 26
1159 Copenhagen
DENMARK

855
Organisationen for Vedvarende Energi
Willemoesgade 14 KLD
2100 Copenhagen
DENMARK

856
Roskilde Universitet
Senteret
4000 Roskilde
DENMARK

857
SPARCO
A/S Naesbjerg Maskinsenter
6800 Varde
DENMARK

858
Teknologisk Institut
Gregersensvej
DK-2630 Tastrup
DENMARK

859
Tvindskilerne
Box 10
6900 Ulfborg
DENMARK

860
UNIMAY
De Forenede Maskinfabrikerne
Nakskov
Tolderlundsvej 2
5000 Odense
DENMARK

861
Vaerlose Gruppen
Christians Minde Vej 11
2100 KBH Ø
DENMARK

862
Vester Hassing Strand
Alborg
DENMARK

863
The Zero-Energy House
The Thermal Insulation Laboratory
Technical University of Denmark
2800 Lyngby
DENMARK

FINLAND

864
BIOS r.y.
Mannerheim 56 B 15
SF-0026 Helsinki 26
FINLAND

FRANCE

865
Action Specifique Energie Solaire
Centre National de la Recherche Scientifique
27 rue Paulbert
94200 Iury
FRANCE

866
Alternative et Technologie
BP 51
75861 Paris Cedex 18
FRANCE

867
Association des Amis des Moulins a Vent
Musee des Arts et des Traditions Populaires
6 route du Mahatma Gandhi

75016 Paris
FRANCE

868
Association Francaise pour l'Etude et le Developpement des Applications de l'Energie Solaire
28 rue de la Source
75016 Paris
FRANCE

869
Compagnie d'Etudes Industrielles et d'Amenagement du Territoire (CINAM)
3, place des Victoires
75001, Paris
FRANCE

870
Compagnie Internationale de Developpement Regional (CIDR)
3.33 rue Marbeuf
Paris 75008
FRANCE

871
Centre d'Exchanges et Promotion des Artisans en Zones a Equiper (CEPAZ)
60 Ave. Philippe-Auguste
75011 Paris
FRANCE

872
Centre d'Etudes et d'Experimentation du Machinisme Agricole Tropical (CEEMAT)
Parc de Tourvoie
92 Antony
Hauts de Seine
FRANCE

873
Centre International de Recherche sur l'Environnement et le Developpement (CIRED)
54, Boulevard Raspail
75270 Paris Cedex 06
FRANCE

874
Centre International pour le Developpement Agricole (CIRDA)
19 Rue Dufrenoy
75116 Paris
FRANCE

875
Centre National de Recherche Scientifique (CNRS)
Solar Energy Laboratory
Odeillo, Pyrenees
FRANCE

876
Cooperation Mediterraneenne pour l'Energie Solaire (COMPLES)
32 Course Pierre Paget
13006 Marseille
FRANCE

877
Delegation aux Energies Nouvelles
13 rue de Bourgogne
75007 Paris
FRANCE

878
Ecodevelopment News
54, Boulevard Raspail, #309
75270 Paris Cedex 06
FRANCE

879
Federation Nationale des Industries des Corps Gras
10 Rue de la Paix
Paris 2
FRANCE

880
Fiches Ecologiques
14 Rue de la Poste
66600 Riversaltes
FRANCE

881
Groupe d'Etude de la Maison Ecologique
84 bs, rue de Grenelle
75207 Paris
FRANCE

882
Groupe de Recherches sur les Techniques Rural (GRET)
34 rue Dumont d'Urville
75116 Paris
FRANCE

883
Institut des Corps Gras
7, Bd De La Tour-Maubourg
Paris 7
FRANCE

884
I.H.R.O. (Institut de Recherches Sur les Huiles et Oleagineux)
9, Squre Petrarque
Paris 16
FRANCE

885
Institut International de Recherche et de Formation (IRFED)
49, rue de la Glaciere
75013 Paris
FRANCE

886
International Federation of Organic Movements
c/o Nature et Progress
3 Chemin de la Bergerie
91700 Ste. Genevieve-des-Bois
FRANCE

887
INODEP
32-34 Avenue Reille
Paris 14
FRANCE

888
Laboratoire de la Roquette
34190 St. Bauzille de Putois
FRANCE

889
Nature et Progres
3 Chemin de la Bergerie F.
Sainte Genevieve des Bois
FRANCE 91700

890
Nature et Vie
13 rue du Village Kervenanec
5600 l'Orient
Bretagne
FRANCE

891
OECD (Organisation for Economic Co-operation and Development)
Development Assistance Directorate
Chateau de la Muette
75 Paris 16
FRANCE

892
OECD (Organisation for Economic Co-operation and Development
Science, Technology and Development
2 rue Andre-Pascal
75775 Paris Cedex 16
FRANCE

893
OECD (Organisation for Economic Co-operation and Development)
Technology and Industrialisation Programme
94 rue Chardon-Lagache
750 Paris Cedex 16
FRANCE

894
OECD
94, rue Chardon-Lagache
75016 Paris
FRANCE

895
Societe d'Aide Technique et de Cooperation (SATEC)
110, rue de l'Universite
Paris
FRANCE

896
Societe Francaise d'Etudes Thermiques et d'Energie Solaire (SOFRETE)
Zone Industrielle d'Amilly
B.P. 163
45203 Montargis
FRANCE

897
UN Educational, Scientific and Cultural Organisation (UNESCO)
7 Place de Fontenoy
75007 Paris
FRANCE

898
Jean-Pierre Girardier
P.D.G. des Establissements Pierre Mengin
320, rue Emile Mengin
B.P. 163 45200
Montargis (Loiret)
FRANCE

GERMANY

899
Aachen Technical University
Research Institute for International Techno-Economic Cooperation
Theatinerstrasse 88
D 5100 Aachen
WEST GERMANY

900
Adapted Technolgies for Developing Countries Working Group
KHG-Zentrum
Nieder-Ramstadter-str. 30
61 Darmstadt
GERMANY

901
BMA Brauschweigische Maschinenbauanstalt
33 Braunschweig
Am Alten Bahnhof 5
Postface 3225
GERMANY

902
Boden und Gesundheit
Postfach 19
D-7183 Langenburg
Wurttemberg
GERMANY

903
Bremen University
28 Bremen 1
Vor dem Steintor 102
WEST GERMANY

904
Consultants for Appropriate Technologies (CAT)
D-7813 Staufen
Etzenbach 16
GERMANY

905
Deutsche Stiftung fur Internationale Entwicklung
Endenicher Str. 41
53 Bonn
WEST GERMANY

906
Freework
6501 Ober-Olm
Obergasse 30
GERMANY

907
Forschungsinstitut fur Internationale Technische-Wietschaftliche Zusammenarbeit
Technischen Hochschule
Vereinsstrasse 3-5
5100 Aachen
GERMANY

908
German Development Assistance for Social Housing Bismarckstrasse 7
D-5000 Koln 1
GERMANY

909
German Foundation for Developing Countries
Zentrale Dokumentation
Endenicherstrasse 41
53, Bonn
GERMANY

910
German Foundation for International Development
ATTN: Mr. Merchert
1 Berlin 27
Reiherwerder
Berlin
GERMANY

911
Interdisciplinary Project Group for Adapted Technologies (IPAT)
Bismark/Onken
Sachtleberstr 37
1000 Berlin
GERMANY

912
Katalyse Technikergruppe
Karl Barth Hous
43 Essen 1
Wittenberg Str 14-16
GERMANY

913
Helmut Milcke
c/o Brot fur die Welt
Stafflenbergstrasse 76
7000 Stuttgart 1
GERMANY

914
Phillips Forschungslaboratorium
Weisshausstrasse
Aachen
GERMANY

915
Prokol Group
Syyelstrasse 46
1 Berlin 12
GERMANY

916
Hans-Christoph Scharpf

Dorfstrasse 18
D-3012 Langenhagen 8
GERMANY

917
Scientific Research Institute for Wind Energy Techniques
Institute of Applied Science
University of Stuttgart
E.U., Pfaffenwaldring 31
D-7000 Stuttgart 80
GERMANY

918
Technische Universitat
Interdisziplinare Projektgruppe fur Angepasste Technologie
Str. des 17 Juni 135
D-1000 Berlin 12
GERMANY

919
Technology Transfer Coordination Centre
c/o IPA
Holzgarternstrasse 17
D-7000 Stuttgart 1
GERMANY

920
Rudiger Ulrich
Trottackerstrasse 2
7880 Sackingen
GERMANY

921
M.F. Ziemek
Planungsgruppe Ritter
Development Consultants
Wiesbadenerstrasse 92
D 6240 Konigstein im Taunus
WEST GERMANY

HUNGARY

922
Research Institute of the Electrical Industry (VKI)
XV. Cservenka Miklos u. 86
P.O. Box 45, Budapest
HUNGARY

ITALY

923
Agrobiologia: Centro Studi di Biologia del Suolo
Via San Stefano 84
40125 Bologna
ITALY

924
FAO-UN Food and Agriculture Organization
Ag. Eng. Services
Via delle Termi de Caracalla
00100 Rome
ITALY

925
Instituto del Chimica Ind.
Universita di Messina
Via Tommaso Cannizzaro
98100 Messina
ITALY

926
Italian Centre for Cooperation with the Building Development of Emerging Nations (C.I.C.S.E.N.E.)
Via Borgosesio 30
10141 Torino
ITALY

927
Montedison
Research and Development Div.
Largo
Doneegani 1/2
20121 Milano
ITALY

928
Save the Earth Project
Molino di Berino
58020 Travale
(GR) ITALY

929
Solar Energy Laboratory
University of Naples
Via F. Crispi 72
Napoli
ITALY

930
Stazione Sperimentale per lo Studio degli Oli, Corpi Grassi
Via G. Columbo
Milano
ITALY

931
Suolo E Salute

48 Via Sacchi
10128 Torino
ITALY

NETHERLANDS

932
Appropriate Technology Group
Mijnbouwplein 11, Kab. 2.15
2628, Delft
NETHERLANDS

933
Central Institute for Industrial Development (CIVI)
Prins Hendrikplein 17
Postbus 1531
The Hague
NETHERLANDS

934
Central Organization for Applied Scientific Research (TNO)
Juliana van Stolberglaan 148
The Hague
NETHERLANDS

935
Centre for Development Planning
Burgemeester Oudlaan 50
Rotterdam 3016
NETHERLANDS

936
Economisch Institut voor het Midden-en Kleinbedrijf
NL's Gravenhage (The Hague)
Neuhuyskade 94
NETHERLANDS

937
Eindhoven University of Technology
Subcommittee for Microprojects
Institute of Development Cooperation
P.O. Box 513
Eindhoven
NETHERLANDS

938
The Energy Centre of the Netherlands
Petten
NETHERLANDS

939
Foundation for the Generation of Electricity by Windmills
Jan Steenlaan 12
Heenstad
NETHERLANDS

940
Intercontinental Education Media
N.V.P.O. Box 42
Aerdenhout
THE NETHERLANDS

941
Netherlands Technical Services
Postbox 5821
2280 HV, Rijswijk Z.H.
NETHERLANDS

942
Projekt de Kleine Aard
Munsel 17
Boxtel, NB
NETHERLANDS

943
RIO Foundation
P.O. Box 299
Rotterdam
NETHERLANDS

944
Tool Stichting Technische Ontwikkelings Landen
Mauritskade 61a
Amsterdam
NETHERLANDS

945
T.W.O.
Personeels Vereniging
Ing. Buro D.H.V.
Laan 1914 35
Amersfoort
NETHERLANDS

946
Technische Hogeschool
Division of Microprojects
Eindhoven Postbus 513
Eindhoven
NETHERLANDS

947
Technische Hogeschool
Steering Committee on Wind Energy in Developing Countries Wind Energy Group
Dept. of Physics
Postbus 513

Eindhoven
NETHERLANDS

948
Technische Physische Dienst TNO
Stieltjerweg 1
Delft PB 155
NETHERLANDS

949
University Wageningen
Landbouwhoeschool
Wageningen
NETHERLANDS

950
Working Group for Development Techniques
Twente University of Technology
P.O.B. 217
Enschede
NETHERLANDS

951
Bureau for International Projects (TNO)
P.O. Box 778
The Hague
NETHERLANDS

NORWAY

952
Institute of Technology
N-7034 Trondheim-NTH
NORWAY

953
Natur og Ungdom
Post Box 8268
Hammersborg
Oslo 1
NORWAY

954
Norwegian Building Research Institute
P.O. Box 322
Blindern
Oslo 3
NORWAY

955
Okoteket i Norge
Torggate 35
Islo 1
NORWAY

956
University of Oslo
P.O. Box 1070
Oslo 3
NORWAY

SPAIN

957
Electrical Industrial Research Association (ASINEL)
Francisco Gervas-3
Madrid 20
SPAIN

958
Hamlet Co-operative
Correos, Cadaques
Prov. Gerona
SPAIN

959
Instituto de Agroquimica y Tecnologia de Alimentos
Jaime Roig II
Valencia 10
SPAIN

960
Instituto de la Gras y sus Derivados
Avenida de Hiliopolis
Sevilla 12
SPAIN

961
Instituto de Informacion y Documentacion en Ciencia y Tecnologia (ICYT)
Joaquin Costa, 22
Madrid 6
SPAIN

962
Junta Nacional Intersindical de Pequena y Mediana Empresa
Paseo del Prado 18 y 20
Madrid 14
SPAIN

SWEDEN

963
Agricultural College of Sweden
Documentation and Information on Agricultural and Rural Development for Developing Countries
The Rural Division Section

Uppsala 7
SWEDEN

964
Beijer Institute
Royal Swedish Academy of Sciences
Stockholm
SWEDEN

965
Chalmers
Development Planning Group
P.O. Box 53156
40015 Gothenburg
SWEDEN

966
Ekoteket
Tjarhousgatan 44
S-11629
Stockholm
SWEDEN

967
Gothenburg Alternative Technology Group
Institute of Theoretical Physics
Chalmers Tekniska Fack
S-40220, Gothenburg
SWEDEN

968
Lund Institute of Technology
Ekologist Byggnade
Dept. of Architecture
P.O. Box 725
S-22-27 Lund
SWEDEN

969
National Swedish Board for Technical Development (STU)
Liljeholmsvage 32
Fack, S-100 72
Stockholm 43
SWEDEN

970
Power Systems Research Group
The Royal Institute of Technology
Stockholm 70
SWEDEN

971
Royal Swedish Academy
Engineering Sciences
Fack 5073
102/42 Stockholm
SWEDEN

972
Svenska Biodynamiska Foreningen
S-15020 Jarna
SWEDEN

973
Swedish Agency for Research Cooperation with Developing Countries (SAREC)
c/o SIDA
S-105 25 Stockholm
SWEDEN

974
Swedish International Development Authority (SIDA)
S-105 25
Stockholm
SWEDEN

975
University of Gothenburg
Botanical Institute
Dept. Microbiology
S-41319 Gothenburg
SWEDEN

976
University of Lund
Division of Physiological Chemistry
Chemical Center
Fack 221-01 Lund 15
SWEDEN

SWITZERLAND

977
Alternative-Katalog
GDI Park im Gurene
CH-8803
SWITZERLAND

978
Bergheimat Gessellschaft
Beaumontweg 11
CH-3007 Bern
SWITZERLAND

979
Biofarm-Genossenschaft Madiswill
CH-4934 Madiswil
SWITZERLAND

980
BlaBla (Blaus Blatt)
Box 97
CH-2900 Porrentruy
SWITZERLAND

981
Buess
Cantonale Landwirtschaftschule
Ebenrain
CH-4450 Sissach
SWITZERLAND

982
CCPD Network Letter
P.O. Box 66
150 Route de Ferney
1211 Geneva
SWITZERLAND

983
Economic Commission for Europe
Environment and Human Settlements
Division
Mr. V. Knoroz
Palais des Nations
1211 Geneva 10
SWITZERLAND

984
Elektro G.M.B.H.
St. Gallerstr. 27
Ch. Wilterthur
SWITZERLAND

985
Environmental Conservation
Elsevier Sequoia SA
P.O. Box 851
CH-1001 Lausanne 1
SWITZERLAND

986
Forschungsinstitut fur Biologischen
Landau
Postfach Oberwil/BL
CH-4104
SWITZERLAND

987
Forum fur Verantwortbare
Anwendung der Wissenschaft
CH 4113 Fluh
SWITZERLAND

988
Latin America Institute
University of St. Gall
Varnbuelstr. 14
9000 St. Gallen
SWITZERLAND

989
Lutheran World Service
P.O. Box 66
Route de Ferney 150
1211 Geneva
SWITZERLAND

990
NIDCS Tractor Project
P.O. Box 450
Manzini
SWITZERLAND

991
Schweizerische Arbeitsgemeinschaft
fur Alternative Technologie (SAGAT)
Postfach 2121
8028 Zurich
SWITZERLAND

992
Swiss Association for Technical Assis-
tance (SATA)
Asylstrasse 41
Postfach 8030
Zurich
SWITZERLAND

993
Swiss Solar Energy Society
Leonhardstr. 27
CH-8001 Zurich
SWITZERLAND

994
Union Internationale de l'Artisanat et
des Petites et Moyennes Entrepri-
ses
98 rue de Saint-Jean
1211 Geneva 11
SWITZERLAND

995
United Nations
Palais des Nations
CH 1211 Geneva 10
SWITZERLAND

996
UN International Labour Organisation
(ILO)
CH-1211 Geneva 22
SWITZERLAND

997
Vitali Kouznetsov
ECE
Palais des Nations
CH 1211 Geneva 10
SWITZERLAND

998
World Council of Churches
Commission on the Churches Participation in Development
Route de Ferney 1211
Geneva 20
SWITZERLAND

999
World Health Organisation
Appropriate Technology for Health Programme
Avenue Appia
1211 Geneva 27
SWITZERLAND

EUROPE—UNITED KINGDOM

1000
Aberdeen College of Education
44 Mid Stocket Road
Aberdeen
UNITED KINGDOM

1001
Alternative Society
9 Morton Ave.
Kidlington, Oxford
UNITED KINGDOM

1002
Alternative Technology Liaison Group (Cumbria)
c/o Barclays Bank Chambers
Crescent Road
Windermere, Cumbria
UNITED KINGDOM

1003
Antipoverty Limited
67 Godstow Road
Wolvercote, Oxford
OV 2 8NY
UNITED KINGDOM

1004
Appropriate Health Resources and Technologies Action Group (AHRTAC-UK)
85, Marylebone High Street
London W1M 3DE
UNITED KINGDOM

1005
Appropriate Technology Yorkshire
Brearton Hall
Brearton, Harrogate
Yorks
UNITED KINGDOM

1006
APV Company Ltd.
P.O. Box 4
Crawley RH10 2QB
England
UNITED KINGDOM

1007
Architecture Assoc.
36 Bedford Sq.
London WC1
UNITED KINGDOM

1008
Architectural Association, Rational Technology Unit
34 Bedford Sq.
London WC1
UNITED KINGDOM

1009
Biodynamic Agricultural Assoc.
Broome Farm, Clent
Stourbridge, Worcestershire
UNITED KINGDOM

1010
Biogas Plant
Easebourne Lane
Midhurst, Sussex GU29 9AZ
UNITED KINGDOM

1011
Biotechnic Research and Development (BRAD)
Eithin-y-Gaer
Church Stoke, Bishop's Castle
Montgomeryshire, Wales
UNITED KINGDOM

1012
Alvan Blanch Development Co., Ltd.
Chelworth, Malmesbury
Wiltshire, SN16 9SG
UNITED KINGDOM

1013
BIT
146 Great Western Road
London W11
UNITED KINGDOM

1014
P. Brachi
23 Gisborne Green

Derby DE1 3NA
UNITED KINGDOM

1015
British Society for Social Responsibility in Science
9 Poland Street
London W1V 3DG
UNITED KINGDOM

1016
Brooksby Agricultural College
Brooksby
Melton Bowbray
Leicestershire, LE14 2IJ
UNITED KINGDOM

1017
Building Research Establishment (BRE)
ATTN: Director
Overseas Division
Bucknalls Lane
Gartston-Watford
Hertford WD2 7JR
England
UNITED KINGDOM

1018
Building Design Partnership
74 Regina Road
Finsbury Park
London N4 3PP
UNITED KINGDOM

1019
Cambridge University Dept. of Architecture
1 Scroope Terrace
Cambridge
UNITED KINGDOM

1020
Centre for Alternatives in Urban Development
ATTN: Dick Kitto
Lower Shaw Farm
Shaw, near Swindon
UNITED KINGDOM

1021
Centre of African Studies
University of Edinburgh
Adam Ferguson Building
Edinburgh EH8 9LL
UNITED KINGDOM

1022
Christian Aid
P.O. Box 1
London SW9 8BH
UNITED KINGDOM

1023
Robin Clark
8 Lambert Street
London
UNITED KINGDOM

1024
Cohen Machinery Ltd.
Woodlane
London W12
UNITED KINGDOM

1025
Commonwealth Agricultural Bureau
Farnham House, Farnham Royal
Slough, Bucks SL2 3BN
UNITED KINGDOM

1026
Commonwealth Bureau of Agricultural Economics
Hartington House
Little Clarendon Street
Oxford OX1 2HH
UNITED KINGDOM

1027
Commonwealth Fund for Technical Cooperation (CFTC)
Commonwealth Secretariat
Marlborough House
Pall Mall, London SW1Y 5HX
UNITED KINGDON

1028
Commonwealth Science Council
Marlborough House
Pall Mall
London SW1Y 5HX
UNITED KINGDOM

1029
Community Technology Group (COMTEX)
13 Bedford St.
Bath, Avon
UNITED KINGDOM

1030
Conservation Tools and Technology, Ltd.
143 Maple Road
Subiton, Surrey KT6 4BH
UNITED KINGDOM

1031
Conservation, Tools and Technology (CTT)
161 Clarence Street
Kingston
Surrey KT 1QT
UNITED KINGDOM

1032
COSIRA
Council for Small Industries in Rural Areas
P.O. Box 717
35 Camp Road
Wimbledon Common SW19 4UP
UNITED KINGDOM

1033
Crofts (Engineers) Ltd.
Thornbury
Bradford 3, Yorks
UNITED KINGDOM

1034
Development Planning Unit
10 Percy Street
London W1P 9FB
UNITED KINGDOM

1035
David Dickson
10 Chaloot Sq.
London NW1
UNITED KINGDOM

1036
Henry Doubleday Research Assoc.
20 Convent Lane
Becking, Braintree
Essex
UNITED KINGDOM

1037
East of Scotland College of Agricultural
West Mains Road
Edinburgh EH9 3J9
UNITED KINGDOM

1038
Ecological Housing Association
11 Lodge End
Radlett, Herts
UNIED KINGDOM

1039
Emerson College
School of Biodynamic Agriculture and Earth Sciences
Forest Row
Sussex
UNITED KINGDOM

1040
Energy Research Group
Open University
Walton Hall
Milton Keynes
Bucks, MK7 AAD
UNITED KINGDOM

1041
Farm and Food Society
4 Willifield Way
london NW11 7XT
UNITED KINGDOM

1042
Fletcher & Stewart Ltd.
Derby
England DE2 8AB
UNITED KINGDOM

1043
Food Production & Rural Development Division
Commonwealth Secretariat
Marlborough House
Pall Mall
London SW1Y 5HX
UNITED KINGDOM

1044
Friends of the Earth (U.K.)
9 Pland Street
London W1V 3DG
England
UNITED KINGDOM

1045
Future Studies Centre
15 Kelso Road
Leeds LS2 9PR
UNITED KINGDOM

1046
Gathering Together
33 Wilson Road
Mount Marrion, County Dublin
Ireland
UNITED KINGDOM

1047
Good Gardeners Assoc.
Arkley Manor
South Herts
UNITED KINGDOM

1048
Hadlow College of Agriculture and Horticulture
Hadlow
Tonbridge, England
UNITED KINGDOM

1049
Hydroponic Advisory and Information Unit
119 Glebe Avenue, Ickenham
Middlesex-England UB10 8PF
UNITED KINGDOM

1050
Imperial College
Dept. of Chemical Eng. and Chemical Technology
Prince Consort Road
London SW7 2BY
England
UNITED KINGDOM

1051
Institute of Child Health
30 Guilford Street
London WC1N 1EH
UNITED KINGDOM

1052
Intermediate Technology Development Group (ITDG)
9 King Street
Covent Garden
London WC2E HN
England
UNITED KINGDOM

1053
International Co-operative Alliance
11 Upper Grosvenor Street
London WIX 9PA
England
UNITED KINGDOM

1054
International Development Centre
Voluntary Committee on Overseas Aid and Development
Parnell House
25 Wilton Road
London SW1V 1JS
UNITED KINGDOM

1055
International Forest Science Consultancy
21 Bigger Road Silverburn
Penicuin EG26 9LQ
Midlothian
UNITED KINGDOM

1056
International Institute for Environment and Development
27 Mortimer Street
London W1A 4QW
UNITED KINGDOM

1057
International Institute of Biological Husbandry
Wye College (University of London)
Wye, Ashford
Kent TN25 5AH
England
UNITED KINGDOM

1058
International Solar Energy Society
The Royal Institute
21 Albermarle St.
London W1X 4BS
UNITED KINGDOM

1059
The International Sugar Journal Ltd.
23A Easton Street
High Wycombe
Bucks, England
UNITED KINGDOM

1060
Intermediate Technology
Publications Ltd.
9 King Street
London WC2E 8HN
UNITED KINGDOM

1061
V. Lamon
64 West Street
South Pethentan
Somerset
UNITED KINGDOM

1062
Laurieston Hall
Laurieston Castle Douglas
Kircudbrightshire, Scotland
UNITED KINGDOM

1063
Leeds University
4 Norwood Grove
Leeds LS6 1DT
UNITED KINGDOM

1064
Loughborough University of Technology
Department of Civil Engineering
Loughborough, Leicestershire
LE11 3TU
UNITED KINGDOM

1065
Low Energy Systems
3 Brighton Road
Dublin 6, Ireland
UNITED KINGDOM

1066
Low Impact Design
Hillside Cottage
Wootton-under-Edge
Glos
UNITED KINGDOM

1067
W. McKinnon & Co., Ltd.
Spring Garden Ironworks
Aberdeen, Scotland
UNITED KINGDOM

1068
Milstead Laboratory of Chemical Enzymology
Broad Oak Road
Sittingbourne
Kent
UNITED KINGDOM

1069
Ministry of Overseas Development
Eland House
Stag Place
London SW1E 5DH
England
UNITED KINGDOM

1070
National Center for Alternative Technology
Llywyngwern Quarry
Powys, Wales
UNITED KINGDOM

1071
National Institute of Agricultural Engineering (NIAE)
Overseas Liaison Department (OLD)
Wrest Part
Silsoe
Bedfordshire MK45 4DT
England
UNITED KINGDOM

1072
The Natural Energy Centre
2 York St.
London W1
UNITED KINGDOM

1073
Newcastle University
Economics Department
Newcastle-upon-Tyne
UNITED KINGDOM

1074
Nigerian Organisation of Women (NOW)
10, Collinson Court
Green Lane
Edgeware, HA8 8BN
UNITED KINGDOM

1075
Open University
Faculty of Technology
Walton Hall, Milton Keynes
MK7 6AA
UNITED KINGDOM

1076
Organic Farmers and Growers Ltd.
Longridge Creeting Road
Stownmarket, Suffolk
UNITED KINGDOM

1077
Oxford Committee for Famine Relief (OXFAM)
Appropriate Technology Officer
274 Banbury Road
Oxford, OX2 7DZ
England
UNITED KINGDOM

1078
Oxford College of Further Education
Anti-Poverty Ltd.
Oxford
UNITED KINGDOM

1079
Oxford University
Dept. of Engineering Science
Parks Road
Oxford
UNITED KINGDOM

1080
Pocklington School
West Green
Pocklington, York YO4 2NJ
UNITED KINGDOM

1081
Portsmouth Polytechnic
The Long House
Owslebury, Mr. Winchester
Hants.
UNITED Kingdom

1082
Pye Research Centre
Walnut Tree Manot, Haughley
Stowmarket, Suffolk
UNITED KINGDOM

1083
Queen's University
Department of Surgery
Institute of Clinical Science
Grosvenor Road
Belfast, Ireland
UNITED KINGDOM

1084
Radtech
71 Thirlwell Road
Sheffield S8 9TF
UNITED KINGDOM

1085
Reading University
APS Department
Whiteknights, Reading
UNITED KINGDOM

1086
Reading University
Engineering Dept.
Whiteknights, Reading
UNITED KINGDOM

1087
The Resource Use Institute Ltd.
"Dunmore"
Pitlochry
Perthshire, PH16 5ED
Scotland
UNITED KINGDOM

1088
Patrick Rivers
Fieldgate, Brockweir
Cepstow NP6 7NN
UNITED KINGDOM

1089
The Rowett Research Inst.
Greenburn Road
Bucksburn, Aberdeen AB2 9SB
England
UNITED KINGDOM

1090
Rural Alternative Project
"New Mills" Luxborough
Watchet, Somerset, TA23 OLF
UNITED KINGDOM

1091
Rural Communications
17 St. James Street
South Petherton
Somerset, England
UNITED KINGDOM

1092
Rural Industries Bureau
35 Camp Road
Wimbledon, London SW19
UNITED KINGDOM

1093
Salford University
St. Barnabas Vicarage
Frederick Road
Salford
UNITED KINGDOM

1094
Shenley Park House
Shenley Church End
Milton Keynes
Bucks, England
UNITED KINGDOM

1095
Society for Environmental Improvement
P.O. Box 11
Godalming, Surrey
UNITED KINGDOM

1096
The Soil Association
Walnut Tree Manor
Houghley, Stowmarket
Suffolk 1P14 3RS
UNITED KINGDOM

1097
Solar Centre Ltd.
176 Ifiled Road
London SW10 9AF
England
UNITED KINGDOM

1098
Street Farm

The Vicarage
21 Flodden Road
London SE5
UNITED KINGDOM

1099
Sussex University
Mantell Building
Falmer Brighton
UNITED KINGDOM

1100
TALC-Teaching Aids at Low Cost
Institue of Child Health
30 Guilford St.
London WC1N 1EH
UNITED KINGDOM

1101
Tin Research Institute
Fraser Road
Greenford, Middlesex
UB6 7AQ
UNITED KINGDOM

1102
Transport and Road Research Laboratory (TRRL)
ATTN: Director
Tropical Section
Browthorne
Berkshire RG11 6AU
England
UNITED KINGDOM

1103
Tropical Products Institute (TPI)
ATTN: Director
56/62 Gray's Inn Road
London WC1X 8LU
England
UNITED KINGDOM

1104
Undercurrents
275 Finchley Rd.
London NW3
England
UNITED KINGDOM

1105
University of Cambridge
Technical Research Division
Dept. of Agriculture
1 Scroope Terrace
Cambridge CB2 1PX
UNITED KINGDOM

1106
University College
Solar Energy Information Office
Dept. of Mechanical Engineering and Energy Studies
Cardiff CF1 1XL
UNITED KINGDOM

1107
University College of Wales, Aberystwyth
Department of Agricultural Economics
Pengalais
Aberystwyth SY23 3DD
UNITED KINGDOM

1108
University of East Anglia
School of Development Studies
Overseas Development Group
Norwich, NR4 7TJ
UNITED KINGDOM

1109
University of Edinburgh
Department of Economics
Edinburgh EH8 9JY
UNITED KINGDOM

1110
University of Edinburgh
Department of Economics
Edinburgh EH8 9JY
UNITED KINGDOM

1111
University of Edinburgh
School of Engineering Science
Kings Building, Mayfield Road
Edinburgh EH9 3JL
UNITED KINGDOM

1112
University of Edinburgh
Department of History
William Robertson Building
Edinburgh EH8 9JY
UNITED KINGDOM

1113
University of Edinburgh
Science Studies Unit
34 Buccleuch Place
Edinburgh EH8 9JY
UNITED KINGDOM

1114
University of Edinburgh

Department of Social Medicine
Usher Institute
Edinburgh EH9 1DW
UNITED KINGDOM

1115
University of Edinburgh
Centre of African Studies
Adam Ferguson Building
Edinburgh EH8 9LL
UNITED KINGDOM

1116
University of Edinburgh
Department of Architecture
18 George Sq.
Edinburgh EH8 9LE
UNITED KINGDOM

1117
University of Glasgow
Department of International Economic Studies
Glasgow, Scotland
UNITED KINGDOM

1118
University of Strathclyde
David Livingstone Institute of Overseas Development
16 Richmond Street
Glasgow G1 1XQ
UNITED KINGDOM

1119
University of Sussex
Institute of Development Studies
Falmer, Brighton BN1 9QH
Sussex, England
UNITED KINGDOM

1120
University of Sussex
International Development Research Centre
Science & Technology Policy Programme
Mantell Building
Falmer, Brighton BN1 9RF
England
UNITED KINGDOM

1121
Voluntary Committee on Overseas Aid and Development (VCOAD)
International Development Centre
Parnell House
25 Wilton Road
London SW1V 1JS
UNITED KINGDOM

1122
War on Want
467 Caledonian Road
London N.7
England
UNITED KINGDOM

1123
Water and Waste Engineering (WEDC)
Dept. of Civil Engineering
University of Technology
Loughborough Leics LE 11 3TU
England
UNITED KINGDOM

1124
World Development Movement
Bedford Chambers
Covent Garden
London WC2E 8HA
UNITED KINGDOM

1125
Wye College
Agrarian Development Unit
Ashford
Kent
UNITED KINGDOM

EUROPE—YUGOSLAVIA

1126
Institute of Mining (RI)
Batajnicki put Broj 2
11081 Zemun
POB 116
Beograd
YUGOSLAVIA

NORTH AFRICA AND NEAR EAST

EGYPT

1127
Ain Shams University
Faculty of Arts
Abbasiyah
Cairo
EGYPT

1128
Alazhar University
Faculty of Agriculture
Food Science Department
Cairo
EGYPT

1129
Alexandria University
Food Science and Technology Department
College of Agriculture
22 Al-Guelsh Avenue
Chatby, Alexandria
EGYPT

1130
Engineering and Industrial Design Development Centre
P.O. Box 2767
Cairo
EGYPT

1131
General Organisation for Industrialisation
4, Moderied E. Tahrir Street
Garden City
Cairo
EGYPT

1132
National Information and Documentation Centre
ATTN: Director
Dokki
Cairo
EGYPT

1133
National Research Centre
Fats and Oil Lab
El-Tahrir Street
Dokki
Cairo
EGYPT

1134
National Research Centre
Office of the Transfer of Technology
El-Tahrir Street
Dokki
Cairo
EGYPT

IRAQ

1135
Specialized Institute for Engineering Industries (SIEI)
P.O. Box 5798
South Gate, Baghdad
IRAQ

ISRAEL

1136
Agricultural Engineering Institute
The Volcani Center
P.O. Box 6
Bet Dagan
ISRAEL

1137
Center for Ground Water Research
The Hebrew University
Jerusalem
ISRAEL

1138
H.L.S., Ltd.
Industrial Engineering Co.
Petah-Tikua
ISRAEL

1139
International Technical Cooperation Centre
P.O. Box 3082
Tel Aviv
ISRAEL

1140
Rural Industrialization Ltd.
P.O. Box 29896
Tel Aviv
ISRAEL

1141
Small Industry Advisory Center
Kiryat Hamelaha
4 Shvil Hameretz
Tel Aviv
ISRAEL

1142
TAHAL Consulting Engineers (Water Planning)
Tel Aviv
ISRAEL

1143
Tel Aviv University
Interdisciplinary Centre for Technological Analysis and Forecasting
Ramat-Aviv
Tel Aviv
ISRAEL

JORDAN

1144
Royal Scientific Society (RSS)
P.O. Box 6945
Amman
JORDAN

KUWAIT

1145
Planning Board
P.O. Box 15
KUWAIT

LEBANON

1146
Conseil National de la Recherche Scientifique
P.O. Box 8281
Beirut
LEBANON

LIBYA

1147
Prof. Y.M. El Sayed
Dept. of Mechanical Engineering
P.O. Box 1098
Tripoli
LIBYA

1148
Industrial Research Centre (IRC)
P.O. Box 3633
Tripoli
LIBYA

MOROCCO

1149
Institut Agronomique and Veterinaire
B.P. 704
Rabat (Agdal)
MOROCCO

OMAN

1150
Khabura Development Project
P.O. Box 9024
Mina al Fahal
OMAN

SUDAN

1151
IBRD
YEI
c/o P.O. Box 913
Khartoum
SUDAN

1152
Industrial Research and Consultancy Institute (IRCI)
P.O. Box 268
Khartoum
SUDAN

1153
International Union for Child Welfare
c/o Department of Youth of the Ministry of Information
P.O. Box 84
Juba
SUDAN

1154
National Council for Research
General Secretariat
P.O. Box 2404
Khartoum
SUDAN

1155
Rural Development Centre
Amadi
SUDAN

1156
Sudan Council of Churches
Committee for Relief & Rehabilitation
P.O. Box 469
Khartoum
SUDAN

1157
Sudan Rural Television Project
P.O. Box 970
Khartoum
SUDAN

1158
University of Khartoum
Department of Mechanical Engineering
Faculty of Engineering and Architecture
Khartoum
SUDAN

1159
University of Khartoum
Institute of Solar Energy & Related Environmental Research
National Council for Research
Khartoum
SUDAN

1160
University of Sudan
Department of Rural Economy
Faculty of Agriculture
Khartoum
SUDAN

TUNISIA

1161
Association pour le Developement de l'Animation Rurale (ASDEAR)
10, rue Eve Nohelle
Tunis
TUNISIA

1162
CARE/MEDICO
18, Avenue Docteur Conseil
Tunis
TUNISIA

1163
Centre de Recherche du Genie Rural (CRGR)
Route de Soukra
B.P. 10
Ariania
Tunis
TUNISIA

1164
Centre National de l'Informatique (CNI)
6, Rue Behassen Ben Chaabane
Tunis
TUNISIA

LATIN AMERICA AND CARIBBEAN

ARGENTINA

1165
Banco Nacional de Desarrollo
Small Industry Department
25 de Mayo 145
Buenos Aires
ARGENTINA

1166
Centro de Investigacion Documentaria
Sra. MC Hepburn de Santacapita
Casilla de Correo 1359
1000 Buenos Aires
ARGENTINA

1167
CONFAGUA
Casilla de Correo 3744
Buenos Aires
ARGENTINA

1168
Instituto Nacional de Tecnologia Industrial (INTI)
Libertad 1235
Buenos Aires
ARGENTINA

1169
INTI - National Institute of Industrial Technology
Casilla de Correo No. 157
Sucursal San Martin
Buenos Aires
ARGENTINA

BARBADOS

1170
Appropriate Technology Resource Service (CADEC)
Christian Action for Development in the Caribbean
P.O. Box 616
Bridgetown
BARBADOS

1171
Deputy Director
Peace Corps/Eastern Caribbean
The Grott, Beckles Road
P.O. Box 696C
Bridgetown
BARBADOS

1172
Caribbean Development Bank
Technology Information Unit
P.O. Box 408
Wildey, St. Michael
BARBADOS

1173
Associate Director
Peace Corps/Eastern Caribbean
P.O. Box 696C
Bridgetown
BARBADOS

BOLIVIA

1174
Agrobiologica
Centro Pedagogico y Cultural de Portales
Casilla 544
Cochobamba
BOLIVIA

1175
Camara Nactional de Industrias
Casilla 611

La Paz
BOLIVIA

1176
Centro Boliviano de Productividad Industrial
Avenida Camcho 1485
La Paz
BOLIVIA

1177
Centro de Investigacion y Promocion del Campesinado (CIPA)
Oruro
BOLIVIA

1178
Centro para el Desarrollo Social y Economico (DESEC)
Casilla 1420
Cochambamba
BOLIVIA

1179
Consortium for International Development
Casilla 8019
La Paz
BOLIVIA

1180
Corporacion Boliviana de Fomento
Casilla de Correo 1124
La Paz
BOLIVIA

1181
Direccion General de Normas y Tecnologia (DGNT)
Jefe, Servicio de Informacion
Tecnica Industrial
Casilla 4430
La Paz
BOLIVIA

1182
Industrial Bank
Casilla 3077
Santa Cruz
BOLIVIA

1183
National Community Development Service
Government Offices
La Paz
BOLIVIA

1184
Research Institute of Mining and Metallurgy (IIMM)
Calle Junin y Petot, Oruro
Casilla 600, Oruro
BOLIVIA

1185
Voluntarios en Accion (VEA)
Apartado Postal 3556
La Paz
BOLIVIA

BRAZIL

1186
FESSC - Fundacao Educacional do Sul de Santa Catarina
Caixa Postal 370
Tubarao, Santa Caterina
BRAZIL

1187
Program Director
Catholic Relief Services USCC (CRS)
Rua Dona Maria Cesar, 170 S/303
50.000 Recife (PERNAMBUCO)
BRAZIL

1188
Fundacao Carlos Alberto Vanzolini
Edificio J.O. Monteiro de Camargo
Cidade Universitaria - CEP 05508
Sao Paulo - S.P.
BRAZIL

1189
Fundacao de Ciencia e Tecnologia (CIENTEC)
R. Washington Luiz 675, Caixa Postal 1864
Porto Alegre, Rio Grande do Sul
BRAZIL

1190
Fundacao de Desenvolvimento da Pesquisa
Av. Antonio Carlos, 6627
Belo Horizonte, Minas Gerais
BRAZIL

1191
Institute of Food Technology (ITAL)
Av. Brasil, 2880
13.100 Campinas
P.O. Box 139
Sao Paulo
BRAZIL

1192
Instituto de Pesquisas Tecnologicas
Divisao de Quimica e Engenharia Quimica
Cidade Universitaria Armando de Salles Oliveira
Caixo Postal 7141
Sao Paulo
BRAZIL

1193
Instituto Nacional de Tecnologia (INT)
Av. Venezuela 82
Rio de Janeiro, CEP-20.000
G.B.
BRAZIL

1194
Laboratorio de Energia Solar
Cidade Universitaria
Joao Pessoa PB
5800
BRAZIL

1195
National Institute of Technology (INT)
Av. Venezuela 86
4 Andar, Guanabara
Rio de Janeiro
BRAZIL

1196
Programa de Desenvolvimento de Micro-Empresas do Reconcavo (PRODEMER)-Jose W. Queiroz Pinto
Rua Ana Nery, 25
Cachoeira
BRAZIL

1197
Programa de Investimento em Micro Empresas
Rua Joaquim Tavora, 27
Santo Antonio
Salvador (BAHIA)
BRAZIL

1198
AITEC/BRAZIL
Rua General Cristovao Barcelos 25
Laranjeiras
Rio de Janeiro (GB)
BRAZIL

1199
Banco Brascan de Investimento S.A.
Gerente Regional
Av. Guararapes, 11-40 Andar
Recife (PERNAMBUCO)
BRAZIL

1200
Banco Nacional de Desenvolvimento Economico
Av. Rio Branco 53
Rio de Janeiro
BRAZIL

1201
Banco de Nordeste de Brasil
Small Industry Division
Industrial & Investment Dept.
Rua Major Facundo No. 500
Fortaleza, Ceara
BRAZIL

1202
Barroslearn
Rua 24 de Maio 62-5 Andar
01041 Sao Paulo S.P.
BRAZIL

1203
Prof. Francisco Javier Barturen
Rua Aristides Novis, 101
Federacao
40.000 Salvador (BAHIA)
BRAZIL

1204
Centro de Pesquisa e Desenvolvimento (CEPED)
Rua Rio Sao Francisco, 1
Salvador, Bahia
BRAZIL

1205
Centro Tecnologico de Minas Gerais (CETEC)
Av. Bias Fortes 401
Caixa Postal 2306
Belo Horizonte, M.G.
BRAZIL

1206
M. Dedni S.A. Metalurgica
Av. Mario Dedini 201
Piracicaba 13400 (SP)
BRAZIL

1207
Irma Dulce
Albergue Santo Antonio
2 Av. Bomfim

Salvador (BAHIA)
BRAZIL

1208
Electrical Power Research Centre (CEPEL)
Cidade Universitaria
Ilha do Fundao
P.O. Box 2754
20000 Rio de Janeiro
BRAZIL

1209
Executive Commission for the Economic Recuperation of Cacao Region (CEPLAC)
KM 26 Radovia Ilheus-Itabuna
Itabuan, Bahia
BRAZIL

1210
Faculty of Food Technology
State University of Campinas
Caixal Postal 1170, 13100
Campinas, Sao Paulo
BRAZIL

1211
Federacao de Orgaos para Assistencia Social e Educacional (FASE)
Rua Pacifico dos Santos, 110
Recife (PERNAMBUCO)
BRAZIL

1212
William Sears Reese
Director Regional, Corpor da Paz
Rua Amaury de Medeiros, 140
Derby
Recife (PERNAMBUCO)
BRAZIL

1213
Paulo Rocha
Assistant Peace Corps Director
Eastern Region/Brazil
Boulevard Suico No. 136 - Nazare
40.000 Salvador (BAHIA)
BRAZIL

1214
Rockefeller Foundation
Rue Des. Balduino Andrade, 60-E
Chame Chame
Caixa Postal 511
40.000 Salvador (BAHIA)
BRAZIL

1215
Secretaria de Economia e Planejamento
Sao Paulo
BRAZIL

1216
Servicio de Voluntarios Alemaes
Av. Conde de Boa Vista, 50
Ed. Pessoa de Melo - 40
50.000 Recife (PERNAMBUCO)
BRAZIL

1217
Servicio Publico Federal
Ministerio do Interior
Superintendencia do Des Envolvimento do Nordeste (SUDENE)
Recife (PERNAMBUCO)
BRAZIL

1218
Superintendencia do Desenvolvimento do Nordeste
Ministerio de Interior
Recife (PERNAMBUCO)
BRAZIL

1219
Technological Center of Minas Gerais (CETEC)
Av. Bias Fortes 401
30.000 Belo Horizonte
Minas Gerais
BRAZIL

1220
Technological Research Institute
Cidado Universitaria
P.O. Box 7141
Sao Paulo
BRAZIL

1221
Uniao Nordestina de Assistencia a Pequenas Organizacoes (UNO)
Rua Gervasio Pires No. 804
Boa Vista
Recife (PERNAMBUCO)
BRAZIL

CHILE

1222
Chilean Institute of Technology (INTEC)
Av. Santa Maria 06500
Casilla 667, Santiago
CHILE

1223
Confederacion Nacional Unica de la Pequena Industria y Artesanado
Huerfanos 1373, Of. 403
Santiago
CHILE

1224
Corporacion Industrial Para el Desarrollo del Area Metropolitana (CIDEME)
Merced 136, Oficina 31
Santiago
CHILE

1225
Corporacion Industrial Para el Desarrollo Regional (CIDERE)
Anibal Quinto 372
Concepcion
CHILE

1226
Estudio Elaboracion Ejecucion (ETRES)
Santo Domingo 573, Of. 85
CHILE

1227
Instituto de Investigaciones Tecnologicas (INTEC/CORFO)
Jefa de Adquisiciones
Casilla 667
Av. Santa Maria 06550
Santiago
CHILE

1228
National Council for Scientific and Technological Research
Canada 308
Casilla 297-V
Santiago
CHILE

1229
Servicio de Cooperacion Tecnica
Casilla 276
Santiago
CHILE

1230
Thermodynamics Laboratory
Universidad Tecnica Federico Santa Maria
Casillas 110-V 132-V
Valparaiso
CHILE

1231
United Nations Economic Commission for Latin America
Latin American Institute of Economic and Social Planning
Casilla Postal 1567
Santiago
CHILE

COLOMBIA

1232
ACOPI - Asociacion Columbiana Popular de Industriales
Apartado Aereo 16451
Bogota
COLUMBIA

1233
ANDI - Asociacion Nacional de Industriales
Apartado Aereo 4430
Bogota
COLOMBIA

1234
Asociacion Nacional de Instituciones Financieras (ANIF)
Service of Assistance to Small and Medium Industry
Calle 12 #7-14, Piso 7
Bogota
COLOMBIA

1235
Centro de Desarrollo Integrado e Investigaciones Tropicales "Las Gaviotas"
Bogota
COLOMBIA

1236
Centro de Desarrollo Integrado "Las Gaviotas"
Ap. Aereo 18261
Bogota
COLOMBIA

1237
Centro de Investigaciones Multidisciplinarias en Desarrollo Rural (C.I.M.D.)
A. Postal No. 3708
Universidad del Valle, Medicina Social
Cali
COLOMBIA

1238
Centro de Investigaciones Multidisciplinarias en Tecnologica y Empleo (CIMTE)
Ap. Aereo 2188
Universidad del Valle, Div. Ingeneria
Cali
COLOMBIA

1239
Centro Internacional de Agricultura Tropical (CIAT)
Ap. Aereo 67-13
Cali
COLOMBIA

1240
Centro Internacional de Investigaciones para el Desarrollo
Oficina Regional America Latina
Apartado Aereo 53016
Bogota
COLOMBIA

1242
C.I.M.T.E.
Apo. Aereo 2188
Cali
COLOMBIA

1243
CIPE - Centro Interamericano de Promocion de Exportaciones (OAS)
Apartado Aereo 5609
Bogota
COLOMBIA

1244
Colombian Rural Reconstruction Movement
Carrera 13, No. 73-59 Of. 201
Apo. Aereo No. 22565
Bogota
COLOMBIA

1245
Corporacion Andina de Fomento
Edificio El Museo
Calle 29 No. 6-58, Of. 703
Bogota
COLOMBIA

1246
Corporacion Financiera Popular
Apo. Aereo 5179
Bogota
COLOMBIA

1247
FUNDAEC
c/o Fundacion Rockefeller
Apo. Aereo 6555
Cali
COLOMBIA

1248
Fundacion Colombiana de Desarrollo
Apo. Aereo 29853
Bogota
COLOMBIA

1249
Fundacion Investigacion de Vivienda
A Aero 50604
Medellin
COLOMBIA

1250
Fundacion para el Fomento de la Investigacion
Cientifica y Technologica
(FICITEC)
ATTN: Manuel A. Botero B.
Apartado Aereo 27872
Bogota
COLOMBIA

1251
Futuro Para La Ninez
Apdo. Postal No. 4445
Medellin
COLOMBIA

1252
J.T. Glauser
Universidad de Los Andes
Apo. Aereo 4976
Bogota
COLOMBIA

1253
Hogares Juvenile Campesinos (H.J.C.)
Calle 56 C N. 50-30
Piso 4
Medellin
COLOMBIA

1254
Industrial Development Program of Banco Popular
Apt. Aereo 1869
Cali
COLOMBIA

1255
Instituto de Fomento Industrial

Ap. Aereo 4222
Bogota
COLOMBIA

1256
Instituto de Investigaciones Tecnologicas (ITT)
Av. 30 No. 52A77
Ap. Aereo 7031
Bogota
COLOMBIA

1257
Instituto Agricola
ATTN: Director
Robles (Valle)
COLOMBIA

1258
International Technology Inc. (INTEC)
Ap. Aereo 3061
Bogota, E.E.
COLOMBIA

1259
Metalibec, S.A.
A A 11798
Bogota
COLOMBIA

1260
Plan Nacional de Alimentacion y Nutricion
Consultant, Appropriate Technology
Cali
COLOMBIA

1261
Save the Children
ATTN: Director
Kra 7 #17064 of 606
Bogota
COLOMBIA

1262
SENA - Servicio Nacional de Aprendizaje
Asesoria a las Empresas
Ap. Aereo 9801
Bogota
COLOMBIA

1263
Cervicio Nacional de Aprendizaje
Ministry of Labor and Social Security
Ap. Aereo 9801
Bogota
COLOMBIA

1264
Universidad de Los Andes
Center for Economic Development Studies
Calle 18 Con Carrera 1-E
Bogota
COLOMBIA

1265
Universidad de los Andes
Facultad de Ingenieria
Ap. Aereo 4878
Bogota
COLOMBIA

1266
Universidad Nacional de Colombia
CID - Development Research Center
Bogota
COLOMBIA

1267
Leonor Uribe de Villegas
Abogada
Av. 5a. Norte No. 23-D-48
Cali
COLOMBIA

1268
Fondo Colombiano de Investigaciones Cientificas (COLCIENCIAS)
Transversal 9, No. 133-28
Ap. Aereo 051
Bogota
COLOMBIA

COSTA RICA

1269
Carozo Ltda.
Ap. 2297
San Jose
COSTA RICA

1270
CATIE
Centro Agronomico Tropical de Investigacion y Ensenanze
ATTN: Humberto Jimenez SAA
CATIE
Turrialba
COSTA RICA

1271
Centro de Investigacion y Capacitacion

RL
P.O. Box 2753
San Jose
COSTA RICA

1272
Centro de Investigaciones en Technologia de Alimentos
Universidad de Costa Rica
San Jose
COSTA RICA

1273
CONICIT - National Council for Scientific and Technological Research
Ap. 10318
San Jose
COSTA RICA

1274
Food Technology Research Center
Universidad de Costa Rica
San Jose
COSTA RICA

1275
Instituto de Fomento y Asesoria Municipal (IFAM)
Calle 1, Av 3
San Jose
COSTA RICA

1276
Instituto Tecnologico de Costa Rica
Centro de Informacion Tecnologica
Ap. 159
Cartago
COSTA RICA

1277
Superior Institute for Rural Training (ISADE)
Ap. Postal 2753
San Jose
COSTA RICA

CUBA

1278
Centro de Investigacion y Experimentacion de la Construccion
Ap. 6180
La Habana
CUBA

DOMINICA

1279
Agricultural and Industrial Development Bank
No. 3 Jewel St.
Roseau
DOMINICA

1280
National Commercial & Dev. Bank
(AID Bank Subsidiary)
64 Hillsborough St.
P.O. Box 215
Roseau
DOMINICA

1281
Mr. Tony Inglis
Dawn Creations
46 Hanover St.
Roseau
DOMINICA

DOMINICAN REPUBLIC

1282
Fundacion Dominicana de Desarrollo
Ap. 85F
Santo Domingo, Z P. No. 1
DOMINICAN REPUBLIC

1283
Fundacion Dominicana de Desarrollo
Calle Mercedes, Apt. 4
Ap. Postal 857
SANTO DOMINGO

1284
Instituto Dominicano de Tecnolgia Industrial (INDOTEC)
ATTN: Deputy Director
A. Jose Nunez de Caceres
Ap. 320-2
SANTO DOMINGO

1285
VITA Dominicana
Ap. 227-2
Centro de los Heros
SANTO DOMINGO

ECUADOR

1286
ASDELA (Asesoria para Desarrollo en

Latino America)
P.O. Box 498
Quito
ECUADOR

1287
Banco Nacional de Fomento
Ap. 685
Quito
ECUADOR

1288
CENAPIA - Centro Nacional de Promocion de la Pequena Industria y Artesania
Casilla Postal 2283
Quito
ECUADOR

1289
Franklin Canelos
Brethren Unidos
Av. Americas 2207
Quito
ECUADOR

1290
Centro de Capacitacion Integral del Campesino de Pastaza
c/o Guillermo Prentice
Casilla 757
Puvo, Oriente
ECUADOR

1291
Centro de Desarrollo Industrial del Ecuador (CENDES)
Jefe, Servicio de Informacion Tecnica
Ap. 5833
Guayaguil
ECUADOR

1292
Centro Ecuatoriano de Servicios Agricolas (CESA)
Mallorca 727
Quito
ECUADOR

1293
Fondo Ecuatoriana Populorum Progressio (FEPP)
Casilla 5202, C.C.I.
Quito
ECUADOR

1294
Fundacion Ecuatoriana de Desarrollo
Ave. Colon #1120 y Juan L. Mera
Ap. Postal 2529
Quito
ECUADOR

1295
Institute for Technological Investigation of the National Polytechnical College
Calle Isable la Catolica
Ap. 2759
Quito
ECUADOR

1296
Instituto Nacional de Investigacion Agropecuaria
Urdesa Las Lomas 417 y 5
Guayaquil
ECUADOR

1297
Ministry of Industry, Commerce and Integration
Program of Credit and Technical Assistance to Small Industry
Ap. Postal 194-A
Quito
ECUADOR

1298
Peace Corps Small Business Program
c/o American Consulate
Guayaquil
ECUADOR

1299
Programa de Promocion de Empresas Agropecuarias (PPEA)
Guayquil
ECUADOR

EL SALVADOR

1300
Centro Nacional de Capacitacion
AGROPECUARIA
ATTN: Director
Blvd. Los Heroes 12-12
San Salvador
EL SALVADOR

1301
Centro Nactional de Productividad (CENAP)
Servicio de Informacion y Transferencia de Tecnologia

Av. Espana 732
San Salvador
EL SALVADOR

1302
Fundacion Salvadorena de Desarrollo y Vivienda
18 Av. Norte 633
Ap. Postal (06) 421
San Salvador
EL SALVADOR

1303
Instituto Salvadoreno de Fomento Industrial
Calle Ruben Dario 628
San Salvador
EL SALVADOR

1304
Universidad Centro-Americana
Jose Simeon Canas (UCA)
Ap. Postal (01) 168
San Salvador
EL SALVADOR

1305
VITA El Salvador
Ap. Postal 421
San Salvador
EL SALVADOR

GUATEMALA

1306
Associacion Guatemateca de la Energia Solar (AGDES)
Calzada Auilar Batres 19-57 Z. 12
Guatemala City
GUATEMALA

1307
Banco Metropolitano, S.A.
ATTN: Director-Gerente
5 A. Avenida 8-24, Zona 1
Guatemale City
GUATEMALA

1308
Bank of America
ATTN: Regional Vice President
P.O. Box 2070
Guatemala City
GUATEMALA

1309
Dr. Carroll Behrhorst
Clinica Behrhorst
Ap. Postal #7
Chimaltenango
GUATEMALA

1310
Hernan Juan Berducido
Gerente, Fundacion del Centavo
8a. Calle 5-09, Zona 9
Guatemala City
GUATEMALA

1311
Centro de Estudios Meso Americanos para la Technologia Apropiada (CEMAT)
ATTN: Director
Ap. Postal 1160
80 Calle 6-06, Zona 1
Guatemala City
GUATEMALA

1312
Centro de Experimentacion en Tecnologia Apropiada
30 Calle 18-20, Zona 12
Guatemala City
GUATEMALA

1313
Centro de Productividad y Desarrollo Industrial
Guatemala City
GUATEMALA

1314
Centro Mesoamericano para Estudias de Tecnologia Apropiada (CEMAT)
8a Calle 6-06, Zona 1
Ap. Postal 1160
Guatemala City
GUATEMALA

1315
Diocese of New Uhm
San Lucas Toliman
Solola
GUATEMALA

1316
Estacion Experimental Choqui
Ap. Postal 159
Quezaltenango
GUATEMALA

1317
Instituto Centroamericano de Investigacion y Tecnologia (ICAITI)

Av. La Reforma 4-47
P.O. Box 1552, Zona 10
Guatemala City
GUATEMALA

1318
Instituto Tecnico de Capacitacion (INTECAP)
Ap. Postal 2568
Guatemala City
GUATEMALA

1319
LAAD de Centroamerica, S.A.
ATTN: Pres.
12 Calle 1-48, Zona 10
Ap. Postal 110-A
Guatemala City
GUATEMALA

1320
Laboratorio de Energia Solar
Estacion Experimental
Canton Chogur
Quezaltenango
GUATEMALA

1321
Director
Save the Children Alliance
Ap. Postal 2903
3a. Calle 7-28, Zona 10
Guatemala City
GUATEMALA

1322
Producciones Carlos Campesino (PCC)
Ap. 2444
Guatemala City
GUATEMALA

1323
B.W. "Pops" Rudder
Executive Manager
American Chamber of Commerce of Guatemala
2d Flr. Room H
9a. Calle 5-54, Zona 1
Guatemala City
GUATEMALA

1324
John Guy Smith
Finca Hawaii
Teculutan
Zacapa
GUATEMALA

1325
Javier Tessari
LAAD de Centroamerica, S.A.
12 Calle 1-48, Zona 10
Ap. Postal 110-A
Guatemala City
GUATEMALA

GUYANA

1326
Small Industry Corp.
229 South St.
Lacytown
Georgetown
GUYANA

1327
Small Industries Corp.
c/o Ministry of Economic Development
Georgetown
GUYANA

HAITI

1328
Rudolph H. Boulos
Directeur Pharmacien
Les Laboratoires "PHARVAL"
45, Rue Traversiere
Boite Postale W-66
Port au prince
HAITI

1329
Lucien Cantave
Directeur-Coordonateur
Service d'Organization de la Vie Rurale (SOVIR)
4 Ave. Ch. Sumner Turgeau
Port au Prince
HAITI

1330
Guy B. Madhere
Comptable
P.O. Box 31
Port au Prince
HAITI

1331
Mr. Jacques Lorthe, Eng.
Office of Science & Technology
CONADEP
Palais des Ministeres
Port au Prince
HAITI

1332
Lionel Laraque
Haiti Water Supply S.A.
Bepite Postale 2075
Port au Prince
HAITI

1333
Institut de Technologie et d'Animation
ITECA
P.O. Box 510
Port au Prince
HAITI

1334
Julien Lauture
Secretaire de la Chambre de Commerce d'Haiti
P.O. Box 982
Port au Prince
HAITI

1335
Michael Madsen
Brasserie Nationale d'Haiti S.A.
P.O. Box 1334
Port au Prince
HAITI

1336
ONAAC - L'Office National de Alphabetization et Action Communautaire
ATTN: Directeur Division de Action Communautaire
P.O. Box 100,000 INT
Port au Prince
HAITI

HONDURAS

1337
Don Anderson
Agency for International Development
U.S. Embassy
Tegucigalpa
HONDURAS

1338
Asociacion Nacional de Industriales de Honduras
Ap. Postal 46
Tegucigalpa
HONDURAS

1339
Centro Cooperative Tecnico Industrial (CCTI)
Av. La Paz 407
Ap. Postal 938
Tegucigalpa
HONDURAS

1340
Centro de Informacion Industrial (CII)
Universidad Nacional Autonoma de Honduras
Tegucigalpa
HONDURAS

1341
Concejal
Concejo Metropolitano del Distrito Central
Tegucigalpa
HONDURAS

1342
Alejandro Corpeno
Save the Children Federation, Inc.
Edificio La Urbana
Segundo Piso
Ap. Postal No. 333
Tegucigalpa
HONDURAS

1343
William G. Nickol
Brown & Root
Edificio Cantero #400
Tegucigalpa
HONDURAS

1344
VITA Honduras
Ap. Postal 786
Altos de la Urbana
Tegucigalpa
HONDURAS

JAMAICA

1345
Dr. R. Gonzales
Scientific Research Council
9 Montery Drive
Kingston 6
JAMAICA

1346
Jamaica Industrial Development Corp.
Small Industries Division

P.O. Box 505
Kingston
JAMAICA

1347
Knox College
ATTN: Rev. Lewis Davidson
Spaldings
JAMAICA

1348
Scientific Research Council
Technical Information Officer
P.O. Box 350
Kingston 6
JAMAICA

MEXICO

1349
Centro de Coordinacion y Promocion Agropecuaria (CECOPA)
Instituto Tecnologico y de Estudios Superiores de Occidente (ITESO)
Av. La Paz 1250
Guadalajara
Estado de Jalisco
MEXICO

1350
Centro de Estudios Economicos y Sociales del Tercer Mundo
A.C., Cori. Porfirio Diaz 50
San Jeronimo Lidice
Mexico 20, D.F.
MEXICO

1351
Centro de Estudios Generales, A.C.
Apartado 732
Chihuahua
Estado de Chihuahua
MEXICO

1352
Centro Internacional de Mejoramiento de Maiz y Trigo (CIMMYT)
Ap. Postal 60641
Mexico, D.F. 6
MEXICO

1353
Centro Nacional de Productividad
Manuel Maria Contreras 133, 2d Floor
Mexico 4, D.F.
MEXICO

1354
CICODET
Ap. M-2007
Mexico 1 D.F.
MEXICO

1355
C.I.M.M.E.T.
(International Maize and Wheat Improvement Center)
Ap. Postal 6-641
Mexico 6, D.F.
MEXICO

1356
Comite Promotor del Desarrollo Socio-Economico del Estado de Nuevo Leon
Cond. del Norte
Desp. 2610-11 y 12
J.I. Ramon 506 Ote.
Monterrey, N.L.
MEXICO

1357
Confederacion de Camaras Industriales de Mexico
Manuel Maria Contreras 133, 8th Floor
Mexico 4, D.F.
MEXICO

1358
Consejo Nacional de Ciencia y Tecnologia (CONACYT)
Codesarrollo Section
Insurgentes Sur 1677 Z 20
Apartado 20-033
Mexico, D.F.
MEXICO

1359
Fundacion Mexicana de Desarrollo
Ejercito Nacional 533
Oficina 602
Mexico 5, D.F.
MEXICO

1360
Jose Giral
Dept. de Ingernieria Quimica
Grupo de Desarrollo de Technologia
Facultad de Quimica
Universidad Nacional Autonoma de Mexico
Mexico, D.F.
MEXICO

1361
INFOTEC-CONACYT

(Information for Industry)
Ap. Postal 19-194
Mexico 19, D.F.
MEXICO

1362
INFYTEC
(Information and Technology)
Apartado 32-0360
Mexico 4, D.F.
MEXICO

1363
Institute of Industrial Research Monterrey Institute of Technology
Av. Tecnologico 2501
Monterrey, N.L.
Sucursal de Correos "J"
Monterrey, N.L.
MEXICO

1364
Instituto de Investigaciones Industriales
Sucursal de Correo "J"
Monterrey, Nuevo Leon
MEXICO

1365
Laboratorios Nacionales de Fomento Industrial
Ap. Postal 410537
Mexico 10, D.F.
MEXICO

1366
Mexican Institute of Technological Research
Calzada Legaria 694
Mexico 10, D.F.
MEXICO

1367
Monterrey Institute of Technology
Department of Special Projects
Sucursal de Correos "J"
Monterrey, Nuevo Leon
MEXICO

1368
Programa de las Naciones Unidas para el Medio Ambiente
President Masaryk 29
Mexico 5, D.F.
MEXICO

1369
Technologica Adecuada
Campeche No. 315
7 Piso
Mexico 11, D.F.
MEXICO

NETHERLANDS ANTILLES

1370
Department for Economic Development
Ansinghstraat No. 1
Willemstad, Curacao
NETHERLAND ANTILLES

1371
Department for Industrialization and Development
Island Government of Curacao
Abraham de Veerstraat 12
Willemstad, Curacao
NETHERLAND ANTILLES

NICARAGUA

1372
Banco Central de Nicaragua
Centro Nicaraguense de Informacion Tecnologia (CENIT)
P.O. Box 2252
Managua
NICARAGUA

1373
Centro de Desarrollo Regional
Proyecto Wisconsin
Puerto Cabezas
Zelaya
NICARAGUA

1374
Ing. Esteban Duque Estrado
Banco Caley-Dagnal
Managua
NICARAGUA

1375
Fundacion Nicaraguense de Desarrollo
Colonia Mantica
Calle Colon
Ap. Postal 2598
Managua
NICARAGUA

1376
Instituto Nicaraguense de Desarrollo
Edificio Inmobiliaria

Managua
NICARAGUA

1377
National Basic Rural Development Project (PLANSAR)
Dept. of Public Health
Managua
NICARAGUA

1378
SEDOC
Servicio de Documentacion y Communicacion Rural
C.E.P.A. Ap. 2929
Managua
NICARAGUA

1379
Small-Scale Handicraft Industries - Nicaragua
c/o U.N. Development Programme
Ap. Postal 3260
Managua
NICARAGUA

1380
Robert Swain
General Manager
Rosario Mining of Nicaragua
Siuna
NICARAGUA

PANAMA

1381
Asociacion Panamena de Desarrollo (APADE)
Calle 34
Edificio Barco de Bogota Apt. 4-404
Panama 9A
PANAMA

1382
Centro para el Desarrollo de la Capacidad Nacional en la Investigacion (CEDECANI)
ATTN: Librarian
Estafeta Universitaria
Universidad de Panama
Panama City
PANAMA

1383
Grupo de Tecnologia Apropriada
P.O. Box 1421
Panama 9A
PANAMA

1384
Rural Development Centre
Apartado 445
Panama 9A
PANAMA

1385
USAID
Paul E. White
Jefe Proyectos Especiales
Ave. Manuel Espinoza B
Edif. Cemento Panama
PANAMA

PARAGUAY

1386
Centro de Desarrollo y Productividad
Asuncion
PARAGUAY

1387
Centro Paraguaya de Documentacion y Information
Asuncion
PARAGUAY

1388
Instituto Paraguayo de Estudios Sociales
Calle Chille 892 1 Piso-Of. 218
Asuncion
PARAGUAY

1389
Instituto Nacional de Tecnologia y Normalizacion
Casilla de Correos No. 967
Asuncion
PARAGUAY

PERU

1390
Banco Industrial del Peru
Technical Assistance Program for Rural Industry
Apartado 1230
Lima
PERU

1391
Gerold Gino Baumann
Embajada Suiza
Casilla 378
Lima
PERU

1392
Edmund H. Benner
InterAmerican Foundation
Representative
1515 Wilson Blvd.
Rosslyn
VA 22209

1393
Centro Internacional de la Papa
Ap. Aereo 5969
Lima
PERU

1394
Comite de la Pequena Industria
Sociedad de Industrias
Casilla 632
Lima
PERU

1395
INDA - Instituto Internacional de Investigacion y Accion Para el Desarrollo (& FONSIAG)
Av. Petit Thouars 3899
San Isidro
Lima 27
PERU

1396
IDINPRO - Instituto Para El Desarrollo Industrial
Ap. 5799
Lima 100
PERU

1397
Instituto de Investigacion Tecnologica Industrial y de Normas Tecnicas (ITINTEC)
Av. Abancay 1176-2 Piso
Lima 1
PERU

1398
Instituto Nacional de Promocion Industrial
Edificio Banco Industrial del Peru
Plaza G. Gastaneta s/n 7 Piso
Lima
PERU

1399
Junta del Acuerdo de Cartagena
Departmento de Industrias
Casilla 3237
Lima
PERU

1400
Novoa Ingenieros Consultores, S.A.
Division Tecnologia
Las Camelias 780
San Isidro, Lima 27
PERU

1401
Projecto Huaylas
Av. La Mar 963
Lima 21
PERU

1402
Universidad Tecnica del Altiplano
Puno
PERU

1403
Ms. Janice M. Weber
Program Analyst
AID c/o U.S. Embassy
Lima 1
PERU

PUERTO RICO

1404
Badrena and Perez
225 Carpenter
Hato Rey
PUERTO RICO 00919

1405
Gurabo Substation
Gurabo
PUERTO RICO

1406
Isabela Substation
Isabela
PUERTO RICO

1407
Mayaguez Institute of Tropical Agriculture
Mayaguez
PUERTO RICO 00708

1408
Centro de Informacion Tecnica
Universidad de Puerto Rico
Recinto Universitario de Mayaguez
Mayaguez
PUERTO RICO

SURINAM

1409
Department of Development
Forest Service
P.O. Box 436
Paramaribo
SURINAM

TRINIDAD

1410
Caribbean Industrial Research Institute (CARIRI)
Information Service to Industry
University Post Office
St. Augustine
TRINIDAD

1411
Industrial Development Corp.
Small Business Division
P.O. Box 949
Port-of-Spain
TRINIDAD

1412
Management Development Centre
Small Business Development Division
P.O. Box 1301
Port-of-Spain
TRINIDAD

1413
Ministry of Housing
48 St. Vincent St.
Port-of-Spain
TRINIDAD

1414
Ministry of Planning and Development
2 Edward St.
Port-of-Spain
TRINIDAD

URUGUAY

1415
Comision Coordinadora para el Desarrollo Economico
Av. Agraciada 1672
Montevideo
URUGUAY

1416
Dr. Patrick Moyna
Missisipi 1634
Ap. 007
Montevideo
URUGUAY

VENEZUELA

1417
Camara de Industriales de Caracas
Ap. 14255
Caracas
VENEZUELA

1418
Comite de Lucha Contra la Desnutricion (COLUDES)
Caracas
VENEZUELA

1419
CONICIT - Consejo Nacional de Investigaciones Cientificas y Tecnologicas
Ap. 70617, Los Ruices
Caracas
VENEZUELA

1420
Corporacion de Desarrollo de la Pequena y Mediana Industria
Maracay
VENEZUELA

1421
Corporacion Venezolana de Fomento
Ap. Aereo 1129
Caracas
VENEZUELA

1422
Federacion de Pequenos y Medianos Industriales de Venezuela
Ap. 1904
Caracas
VENEZUELA

1423
Instituto Venezolano de Investigaciones Cientificas (IVIC)
Caracas
VENEZUELA

1424
Red de Informacion en Ingenieria Arguitectura y Afines (RIIAA)
NTIS/REDINARA
Ap. de Correos 2006

Caracas
VENEZUELA

1425
Universidad de Carabobo
Oficina de Desarrollo Industrial
Apartado 820
Valencia
VENEZUELA

VIRGIN ISLANDS

1426
Island Resources Foundation, Inc.
P.O. Box 4187
St. Thomas 00801
VIRGIN ISLANDS

NORTH AMERICA

CANADA

1427
Association of Geoscientists for International Development (AGID)
c/o Dept. of Geology
Memorial University
St. John's
Newfoundland AIC 5S7
CANADA

1428
Bakavi
P.O. Box 2011, Station D
Ottawa, Ontario
CANADA

1429
Brace Research Institute
P.O. Box 400, MacDonald College of McGill University
Ste. Anne de Bellevue 800
Quebec HOA ICO
CANADA

1430
British Columbia Research Council (B.C. Research)
3650 Westbrook Mall
Vancouver, B.C. V6S 2L2
CANADA

1431
Canada Agricultural Research Station
University Campus
Saskatoon, S7N 0X2
Saskatchewan
CANADA

1432
Canada Department of Agriculture
Research Branch
K.W. Neatby Bldg.
Ottawa, Ontario K1A 0C6
CANADA

1433
Canadian Hunger Foundation
75 Sparks Street
Ottawa, Ontario K1P 5A5
CANADA

1434
Canadian International Development Agency
200, rue Principale
Hull, Quebec K1A 0G4
CANADA

1435
Canadian International Development Agency
Policy Analysis Branch
122 Bank Street
Ottawa, Ontario K1A 0G4
CANADA

1436
Canadian University Service Overseas
151 Slater Street
Ottawa, Ontario K1P 5H5
CANADA

1437
Center for Industrial Research of Quebec (CRIQ)
333 Franguet, Ste-Foy, P.Q.
P.W. GIV 4C7
CANADA

1438
Coady International Institute
St. Francis Xavier University
Antigonish, Nova Scotia
CANADA

1439
Dept. of the Environment
Advanced Concepts Centre
Ottawa
CANADA

1440
Energy Probe
43 Queen's Park Crescent E
Toronto M5S 2C3
CANADA

1441
Food Research Institute
U.S.A. Institute of Food Technologists
Department of Agriculture
Ottawa
CANADA

1442
Industrial Development Bank
P.O. Box 6021
Montreal 101, Quebec
CANADA

1443
Institute for the Study and Application of Integrated Development
17 Inkerman Street
Toronto M4Y 1M5
Ontario
CANADA

1444
International Communications Institute
Box 8268 Station "F"
Edmonton, Alberta T6H 4P1
CANADA

1445
International Development Research Centre (IDRC)
P.O. Box 8500
Ottawa, Ontario K1G 3H9
CANADA

1446
Christian de Laet
P.O. Box 40
Victoria Station
Montreal, Quebec
CANADA

1447
Manitoba Department of Northern Affairs
405 Broadway
Winnipeg, Manitoba R3C 3L6
CANADA

1448
Mennonite Central Committee
210-1483 Pembing Highway
Winnipeg, Manitoba
CANADA

1449
Minimum Cost Housing
School of Architecture
McGill University
P.O. Box 6070
Montreal
CANADA

1450
Mouvement Pour l'Agriculture Biologique
340 Willowdale, Apt. 2
Montreal, Quebec
CANADA

1451
Mylora Organic Farms
960 Apt. 5
Richmond, British Columbia
CANADA

1452
National Research Council of Canada
National Aeronautical Establishment
Montreal Road
Ottawa, Ontario K1A 0R6
CANADA

1453
National Research Council of Canada
Technical Information Service
Ottawa, Ontario K1A 0S3
CANADA

1454
New Brunswick Research and Productivity Council (RPC)
College Hill Road
Frederickton, New Brunswick
P.O. Box 6000
Frederickton, N.B. E3B 5H1
CANADA

1455
North-South Institute
Suite 402
60 Queen
Ottawa K1P 547
CANADA

1456
Edward Pidgeon
25 Picard Place
Kingston, Ontario K7M 2W5
CANADA

1457
Pollution Probe, Ottawa
53 Queen Street, Suite 54
Ottawa, Ontario K1P 5C5
CANADA

1458
Pulp and Paper Research Institute of Canada (PAPRICAN)
570 St. John's Blvd.
Pointe Claire, Quebec H9R 3J9
CANADA

1459
Saskatchewan Research Council
Chemistry Division
30 Campus Drive
Saskatoon, Saskatchewan S7N 0X1
CANADA

1460
United Church of Canada
Division of World Outreach
Room 417, 85 St. Clair Ave. E.
Toronto, Ontario M4T 1M8
CANADA

1461
University of Manitoba
Dept. of Civil Engineering
Winnipeg, R3T 2N2
CANADA

1462
University of Saskatchewan
Animal Science Department
Saskatoon, Saskatchewan S7N 0W0
CANADA

1463
University of Saskatchewan
Department of Biology
Regina Campus
Regina S4S DA2
CANADA

1464
University of Toronto
Faculty of Management Studies
CIDA-NAIROBI Project
246 Bloor Street West
Toronto, Ontario M5S 1V4
CANADA

1465
World Association of Industrial and Technological Research Organisations (WAITRO)
3650 Westbrook Crescent
Vancouver 8
CANADA

1466
Oxfam Quebec
169 St. Paul East
Montreal, P.Q.
CANADA

UNITED STATES—ALASKA

1467
Insulation Engineering, Inc.
1/2 Mile Van Horn Rd.
S R Box 60140
Fairbanks, AK 99701

UNITED STATES—ALABAMA

1468
Alabama Cooperative Extension Service
Duncan Hall
Auburn, AL 36830

1469
Alabama Forestry Commission
513 Madison Ave.
Montgomery, AL 36130

1470
Centre for Aquaculture
Auburn University
Auburn, AL 36830

UNITED STATES—ARKANSAS

1471
Arkansas Community Organizations for Reform Now (ACORN)
523 W. 15th St.
Little Rock, AR 72202

1472
Arrakis Propane Conversions
Route 2, Box 96C
Leslie, AR 72645

1473
Cotton Branch Experiment Station

P.O. Box 522
Marianna, AR 72460

1474
Environmental Design
ATTN: Joel Davidson
Dutton, AR 72726

1475
Helena Cottonoil Mill
ATTN: Joe Brady
Helena, AR 72342

1476
The Ozark Institute
ATTN: Edd Jeffords
Box 549
Eureka Springs, AR 72632

1477
Rice Branch Experiment Station
P.O. Box 351
Stuttgart, AR 72160

1478
University of Arkansas
ATTN: Kenneth S. Bates
Cooperative Extension Service
1201 McAlmont
P.O. Box 391
Little Rock, AR 72203

UNITED STATES—CALIFORNIA

1479
The Aero-Power Company
2398 4th Street
Berkeley, CA 94710

1480
Arcology Circle, Inc.
1932 Foothill Blvd.
Oakland, CA 94606

1481
The Asia Foundation
P.O. Box 3223
San Francisco, CA

1482
Blazing Showers
ATTN: Sundance and Louie
Box 327
Point Arena, CA 95468

1483
Blue Skies Radiant Homes
40819 Park Avenue
Hemet, CA 92343

1484
California Conservation Corps
1530 Capital Ave.
Sacramento, CA 95814

1485
California Cooperative Rice Research Foundation
P.O. Box 306
Biggs, CA 95917

1486
California State College
Department of Economics
Chico, CA 95926

1487
Central Coast Counties Development Corporation
7000 Soquel Dr.
Aptos, CA 95003

1488
Citizens for Energy Conservation and Solar Development, Inc.
P.O. Box 49173
Los Angeles, CA 90049

1489
Cole's Power Models
1355 Church Street
Ventura, CA 93001

1490
Community Environmental Council
109 East de la Guerra
Santa Barbara, CA 93101

1491
Community Technical Services
1932 Foothill Blvd.
Oakland, CA 94606

1492
Earthmind
5246 Boyer Road
Mariposa, CA 95338

1493
Earthwork
1499 Potrero
San Francisco, CA 94110

1494
Ecology Action/Common Ground

2225 El Camino Real
Palo Alto, CA 94306

1495
Ecology Action Educational Institute
P.O. Box 3895
Modesto, CA 95352

1496
Ecology Action of the Midpeninsula
2225 El Camino Real
Palo Alto, CA 94306

1497
Ecology Center of Southern California
P.O. Box 24388
Los Angeles, CA 90024

1498
Experimental Cities, Inc.
P.O. Box 731
Pacific Palisades, CA 90272

1499
Farallones Institute
15290 Coleman Valley Road
Occidental, CA 95465

1500
The Food Bank
Economic and Social Opportunities, Inc.
1460 Koll Circle
San Jose, CA 95112

1501
Farmer's Organic Group
407 Furlong Rd. Dusty Lane
Sebastopol, CA 95472

1502
The Grantsmanship Center
1015 W. Olympic Blvd.
Los Angeles, CA 90015

1503
Harvey Mudd College
The Engineering Clinic
Claremont, CA 19711

1504
Helion, Inc.
Box 4301
Sylmar, CA 91342

1505
George Helmholz
Route 1, Box 24-A
Covelo, CA 95428

1506
Intermediate Technology
Apt. 6
556 Santa Cruz Avenue
Menlo Park, CA 94025

1507
International Association for Education, Development and Distribution of Lesser Known Food Plants and Trees
P.O. Box 599
Lynwood, CA 90262

1508
International Project for Soft Energy Paths
124 Spear Street
San Francisco, CA 94105

1509
Living Systems
Route 1, Box 170
Winters, CA 95694

1510
Meals for Millions
P.O. Box 1666
Santa Monica, CA 90406

1511
Medical Self-Care Magazine
P.O. Box 718
Inverness, CA 94937

1512
National Land for People
1759 Fulton, Room 11
Fresno, CA 93721

1513
New Alchemy Institute West
Box 376
Pascadero, CA 94060

1514
New Ways to Work
457 Kingsley Ave.
Palo Alto, CA 94393

1515
The Nucleus for the Exploration of Humanity's Future
5833 Eucalyptus Lane
Los Angeles, CA 90042

1516
Office of Appropriate Technology
P.O. Box 1677
Sacramento, CA 95808

1517
Bill & Helga Olkowski
1307 Acton St.
Berkeley, CA 94706

1518
Ariel Parkinson
1001 Cragmont Ave.
Berkeley, CA 94708

1519
Portola Institute
558 Santa Cruz Ave.
Menlo Park, CA 94025

1520
Fred Rice Productions, Inc.
P.O. Box 643
La Quinta, CA 92253

1521
Rincon Vitova Insectories, Inc.
P.O. Box 95
Oak View, CA 93022

1522
Round Valley Institute for Man and Nature
Box 493
Covelo, CA 95428

1523
Self-Help Enterprises
220 South Bridge Street
Visalia, CA 93277

1524
Sencenbaugh Wind Electric
2235 Old Middlefield
Mt. View, CA 94040

1525
Project SEW, Inc. (Senior Employment Workshop)
771 Freedom Blvd.
Watsonville, CA 95076

1526
The Sea Group
David Wright, Environmental Architect
P.O. Box 49
The Sea Ranch, CA 95497

1527
Solar Energy Digest
7401 Salerno St.
San Diego, CA 92111

1528
Solare
P.O. Box 4322
Whittier, CA 90607

1529
Southern California Institute of Architecture
1800 Berkeley St.
Santa Monica, CA 90402

1530
Stanford Research Institute (SRI)
333 Ravenswood Avenue
Menlo Park, CA 94025

1531
Survival Subsistence, and Perennial Farming Foundation
225 E. Utah
Fairfield, CA 94533

1532
Terran Sietch Associates
1052 Clark Ave.
Mountain View, CA 94040

1533
The Trust for Public Land
82 Second Street
San Francisco, CA 94105

1534
University of California
Department of Vegetable Crops
Hunt Hall
Davis, CA 95616

1535
University of California
Interdisciplinary Systems Group
Department of Zoology
Davis, CA 95616

1536
University of California at Los Angeles
The Center for the Quality of Working Life
Institute of Industrial Relations
Los Angeles, CA 90025

1537
University of California

Department of Anthropology
Los Angeles, CA 90024

1538
University of California
Sanitary Engineering Research Laboratory
Richmond Field Station
1301 S. 46th Street
Richmond, CA

1539
University of California (Santa Cruz)
International Federation of Appropriate Technology
Department of Environmental Studies and Planning
Santa Cruz, CA 95064

1540
Village Design
1545 Dwight
Berkeley, CA 94703

1541
Volunteers in Asia
Appropriate Technology Project
Box 4543
Stanford, CA 94305

1542
Edward F. Wehlage and Associated Engineers
10707 East Orange Drive
Whittier, CA 90606

1543
West Side Field Station
P.O. Box 158
Five Points, CA 93624

1544
Whole Earth Truck Store
558 Santa Cruz Av.
Menlo Park, CA 94025

1545
Wind Energy Society of America
1700 EAst Walnut St.
Pasadena, CA 91106

UNITED STATES—COLORADO

1546
Bio-Gas of Colorado, Inc.
Research Division
5620 Kendall Court, Unit G
Arvada, CO 80002

1547
Colorado Organic Growers and Marketers Association
2555 West 37 Ave.
Denver, CO 80211

1548
Colorado State University
Engineering Research Center
Fort Collins, CO 80523

1549
Colorado Sunworks
P.O. Box 455
Boulder, CO 80302

1550
Richard L. Crowther
310 Steele Street
Denver, CO 80206

1551
Denver Research Institute
Office of International Programs
University of Denver
University Park
Denver, CO 80208

1552
Domestic Technology Institute
Box 2043
Evergreen, CO 80439

1553
Roaring Fork Resource Center
P.O. Box 9950
Aspen, CO 81611

1554
Solar Technology Corporation
2160 Clay Street
Denver, CO 80211

1555
Suncract Company
Route 4, Box 90
Golden, CO 80401

UNITED STATES—CONNECTICUT

1556
Les Auerback
242 Copse Road
Madison, CT 06443

1557
Community Development Foundation

48 Wilton Road
Westport, CT 06880

1558
Earth Metabolic Design, Inc.
Box 2016, Yale Station
New Haven, CT 06520

1559
Falbel Energy Systems Corporation
472 Westover Road
Stamford, CT 06902

1560
Goldmark Communications Corporation
98 Commerce Road
Stamford, CT 06904

1561
Mr. George Metcalfe
Technoservics
36 Old Kings Hwy., South
Darien, CT 06820

1562
Technoserve
36 Old King's Highway South
Darien, CT 06820

1563
Westport Research Associates
P.O. Box 107
Westport, CT 06880

1564
Yale University
Economic Growth Center
P.O. Box 1987
Yale Station
New Haven, CT 06520

UNITED STATES—DELAWARE

1566
University of Delaware
Community & Resource Development
R.D. 2, Box 48
Georgetown, DC 19947

1567
University of Delaware
College of Marine Studies
Newark, DC 19711

UNITED STATES—DISTRICT OF COLOMBIA

1568
Action/Peace Corps
806 Connecticut Ave.
Washington, DC 20525

1569
American Federation of Labor/Council of Industrial Organizations
Director of International Affairs
815 16th St., N.W.
Washington, DC 20006

1570
Appropriate Technology International
1709 "N" Street N.W.
Washington, DC 20036

1571
Center for Community Change
1000 Wisconsin Avenue, N.W.
Washington, DC 20007

1572
Center for Renewable Resources
Suite 1000
1028 Connecticut Ave., N.W.
Washington, DC 20036

1573
Center for Science in the Public Interest
1757 S. St., N.W.
Washington, DC 20009

1574
Community Technology Inc.
1520 New Hampshire Ave., N.W.
Washington, DC 20036

1575
Congressional Rural Caucus
309 House Office Building
Annex No. 1
Washington, DC 20515

1576
Cooperative League of the USA
International Programs
1828 L Street, N.W.
Room 1100
Washington, DC 20010

1577
Council of the Americas
1700 Pennsylvania Ave., N.W.
Washington, DC 20006

1578
Design Alternatives, Inc.
1312 18th Street, N.W.
Washington, DC 20036

1579
Earth Resources Development Administration
Office of International R&D Programs
20 Massachusetts Avenue, N.W.
Washington, DC 20545

1580
Experience, Inc.
AT Project for Africa
1725 K St., N.W.
Suite 312
Washington, DC 20006

1581
Farmers Crop Service
Cooperative Development Program
U.S. Department of Agriculture
Washington, DC 20250

1582
Farmers Union International
1012 14th Street, N.W.
Washington, DC 20005

1583
Federal Energy Administration (FEA)
National Energy Information Center
Room 1404
12th and Pennsylvania Ave., N.W.
Washington, DC 20461

1584
Foundation Center
1001 Connecticut Ave., N.W.
Washington, DC 20036

1585
Foundation for Cooperative Housing
1001 15th St., N.W.
Washington, DC 20005

1586
Friends of the Earth
620 C St., S.E.
Washington, DC 20003

1587
The Grantsmanship Center
1728 L St., N.W.
Suite 300
Washington, DC 20036

1588
Institute for International Policy
122 Maryland Avenue, N.E.
Washington, DC 20002

1589
The Institute for Local Self-Reliance
1717 18th St., N.W.
Washington, DC 20036

1590
Institute for Policy Studies
1520 New Hampshire Avenue
Washington, DC 20036

1591
Inter-American Development Division
808 17th Street, N.W.
Washington, DC 20577

1592
International Bank for Reconstruction and Development (IBRD)
1818 H. Street, N.W.
Washington, DC 20433

1593
International Cooperative Development Association
1012 14th Street, N.W.
Washington, DC 20005

1594
International Cooperative Housing Development Association
1012 14th Street, N.W.
Washington, DC 20005

1595
International Development Research Center
1028 Connecticut Avenue, N.W.
Washington, DC 20036

1596
International Society for Technology Assessment
Cleveland Park Station
P.O. Box 4926
Washington, DC 20008

1597
International Voluntary Services
1717 Massachusetts Ave., N.W.
Washington, DC 20036

1598
League for International Food Educa-

tion
1126 16th St., N.W. Room 404
Washington, DC 20036

1599
Mid-Atlantic Appropriate Technology Network (MATNET)
1413 K St., N.W.
Washington, DC 20005

1600
National Academy of Sciences
Board on Science and Technology for International Development
2101 Constitution Avenue, N.W.
Washington, DC 20418

1601
National Academy of Sciences
Commission on International Relations
2101 Constitution Ave., N.W.
Washington, DC 20418

1602
National Bureau of Standards
Center for Building Technology
Building 226, Room B-266
Washington, DC 20234

1603
National Bureau of Standards
Office of Energy Related Inventions
Room B-362 Met.
Washington, DC 20234

1604
National Bureau of Standards
Office of International Relations
Washington, DC 20234

1605
National Center for Appropriate Technology
1522 K St., N.W.
Suite 1036
Washington, DC 20005

1606
National Council for International Health
2600 Virginia Ave., N.W.
Suite 600
Washington, DC 20037

1607
National Council for the Public Assessment of Technology
1785 Massachusetts Ave., N.W.
Washington, DC 20036

1608
National Institute of Oilseed Products
1725 K Street, N.W.
Washington, DC 20006

1609
National Rural Electric Cooperative Association
1800 Massachusetts Ave., N.W.
Washington, DC 20036

1610
National Science Foundation
Research Applied to National Needs (RANN)
Division of Exploratory Research and Systems Analysis
Washington, DC 20550

1611
National Science Foundation
Scientists and Engineers in Economic Development
Division of International Programs
Washington, DC 20550

1612
National Science Foundation
Office of International Programs
Washington, DC 20550

1613
National Technical Information Service
Office of the Director
425-13th St., N.W.
Suite 620
Washington, DC 20004

1614
"New Directions"
2021 L St., N.W.
Washington, DC 20036

1615
Organization of American States (OAS)
Technology Transfer Program
1735 I Street, N.W.
Washington, DC 20006

1616
Overseas Development Council
1717 Massachusetts Ave., N.W.
Suite 501
Washington, DC 20036

1617
PADCO, Inc. - Planning and Development Collaborative International
1834 Jefferson Place, N.W.
Washington, DC 20036

1618
Pan American Development Foundation
1725 K Street, Room 1409
Washington, DC 20006

1619
Partners of the Americas
20001 S St., N.W.
Washington, DC 20009

1620
Philippine Sugar Association
1001 Connecticut Ave., N.W.
Washington, DC 20036

1621
Cameron L. Pippitt
Experience, Incorporated
1725 K Street, N.W.
Suite 312
Washington, DC 20006

1622
Research Council for Small Business and the Professions
2120 L Street, N.W.
Washington, DC 20036

1623
Rural Housing Alliance
1346 Connecticut Ave., N.W.
Washington, DC 20036

1624
Society for International Development
1346 Connecticut Ave., N.W.
Washington, DC 20036

1625
Solar Electric International
Dr. Stephen V. Allison
7315 Wisconsin Avenue
Suite 440 W
Washington, DC 20014

1626
Solar Energy Institute of North America
1110 6th St., N.W.
Washington, DC 20001

1627
United States Cane Sugar Refiners Association
1001 Connecticut Ave., N.W.
Room 610
Washington, DC 20036

1628
U.S. Congress
Office of Technology Assessment
Washington, DC 20510

1629
U.S. Department of Agriculture
Economic Research Service
International Development Center
L4th Street and Independence Avenue, S.W.
Washington, DC 20250

1630
U.S. Department of Agriculture
Development Project Management Center
Room 3547
Washington, DC 20250

1631
U.S. Department of Commerce
Economic Development Administration
Local Public Works Program
Washington, DC 20230

1632
U.S. Department of State
U.S. Agency for International Development
Room 2242
Washington, DC 20523

1633
Volunteer Development Corps
1629 K Street, N.W.
Washington, DC 20036

1634
World Future Society
4916 Elmo Ave.
Washington, DC 20014

1635
Worldwatch Institute
1776 Massachusetts Ave., N.W.
Washington, DC 20036

1636
National Academy of Science

National Research Council
2101 Constitution Avenue
Washington, DC 20418

1637
U.S. Department of State
AID
Development Support/Development Information and Utilization
Washington, DC 20523

UNITED STATES—FLORIDA

1638
Agricultural Research and Education Center, Belle Glade
P.O. Drawer A
Belle Glade, FL 33430

1639
Alternative Housing Research Project (AHRP)
6698 N.W. 62nd Ave.
Ocala, FL 32670

1640
American Wind Turbine Co.
Box 446
St. Cloud, FL

1641
Conservation International, Inc.
3069 E. Commercial Blvd.
Fort Lauderdale, FL 33308

1642
Environmental Health and Light Research Institute
3112 Southgate Circle
Sarasota, FL 33579

1643
Environmental Information Centre of the Florida Conservation Foundation, Inc.
935 Orange Ave.
Winter Park, FL 32789

1644
Florida Sugarcane League
P.O. Box 1148
Clewiston, FL 33440

1645
Kinetics Corporation
1121 Lewis Ave.
Sarasota, FL 33577

1646
Progressive Technology Co.
P.O. Box 20049
Tallahassee, FL 32304

1647
Enos L. Schera, Jr.
8254 S.W. 37th Street
Miami, FL 33155

1648
Sun Power Systems, Inc.
1121 Lewis Avenue
Sarasota, FL 33577

1649
University of Florida
Food and Resource Economics Department
Gainesville, FL

UNITED STATES—GEORGIA

1650
Ron Cornman
6640 Akers Mill Rd., N.W.
Apt. 26B 10
Atlanta, GA 30329

1651
Genesis Housing and Community Development Corporation
P.O. Box 827
Claxton, GA 30417

1652
Georgia Institute of Technology
Engineering Experiment Station
Economic Development Laboratory (EDL)
Atlanta, GA

1653
Georgia Institute of Technology
Engineering Experiment Station
International Development Data Center
Atlanta, GA 30332

1654
Georgia State University
Institute of International Business
University Plaza
Atlanta, GA 30303

1655
Rural Development Center

University of Georgia
Tifton, GA 31794

1656
Edith Shedd and Associates, Inc.
Route 2, Box 6111
Monroe, GA 30655

1657
University of Georgia
Agriculture and International Affairs
236 Westview Drive
Athens, GA 30601

UNITED STATES—HAWAII

1658
East-West Technology and Development Institute
East-West Center
1777 East-West Road
Honolulu, HI 96822

1659
Hawaiian Sugar Planters' Association Experiment Station
1527 Keeaumoku Street
Honolulu, HI 96822

1660
International Society of Sugar Cane Technologists
P.O. Box 1057
Alea, HI 96701

1661
Kauai Research Center
RR 1
P.O. Box 278-A
Kapaa, HI 96746

1662
Pacific Science Association
P.O. Box 6037
Honolulu, HI 96818

1663
University of Hawaii
Water Resources Research Center
2500 Campus Road
Honolulu, HI 96822

UNITED STATES—IDAHO

1664
Idaho State University
ATTN: Director
Idaho Statehouse
Boise, ID 83720

1665
University of Idaho
Advanced Building Technology Group
Department of Architecture
Moscow, ID 83843

UNITED STATES—ILLINOIS

1666
Donors' Forum of Chicago
208 S. LaSalle
Chicago, IL 60604

1667
Illinois Institute of Technology
10 West 32nd St.
Chicago, IL 60616

1668
Institute of Food Technologists
221 North La Salle Street
Chicago, IL 60601

1669
Midwest Energy Alternatives Network
Governor State University
Park Forest South, IL 60466

1670
Northern Illinois University
Department of Industry and Technology
DeKalb, IL 60115

1671
Sheaffer & Roland, Inc.
130 N. Franklin St.
Chicago, IL 60606

1672
SOLAG
RR 2
Roseville, IL 61473

1673
University of Chicago Press
Economic Development and Cultural Change (Journal)
5801 Ellis Avenue
Chicago, IL 60637

1674
University of Illinois

Center for Advance Computation
Urbana, IL 61801

1675
Western Illinois University
The Journal of Developing Areas
900 West Adams Street
Macomb, IL 61455

UNITED STATES—INDIANA

1676
American Wind Energy Association
54468 CR 31
Bristol, IN 46507

1677
Ball Corporation
Ball Food Preservation Program
345 High Street
Muncie, IN 47302

1678
Butler University
Department of Economics
Indianapolis, IN 46208

1679
Jester Press
54468 CR 31
Bristol, IN 46507

1680
Purdue University
Department of Agricultural Engineering
W. Lafayette, IN 47907

1681
Purdue University
Department of Chemical Engineering
West Lafayette, IN 47907

1682
Purdue University
Intermediate Technology—Purdue
School of Chemical Engineering
W. Lafayette, IN 47907

1683
Purdue University
Program in Science, Technology and Public Policy
West Lafayette, IN 47906

1684
University of Indiana
PASITAM—Program of Advanced Studies in Institution Building and Technical Assistance Methodology
1005 East 10th Street
Blooming, IN 47401

1685
Wind Power Digest
54468 CR 31
Bristol, IN 46507

UNITED STATES—IOWA

1686
David Brandies Family
Rt. 2
Wilton, IA 52778

1687
Iowa State University
Agricultural Engineering Department
Davidson Hall
Ames, IA 50011

1688
Iowa State University
Technology and Social Change Program
ATTN: Electrical Engineering Department
Ames, IA 50010

1689
New Pioneer Co-operative Society
529 S. Gilbert
Iowa City, IA 52240

1690
University of Northern Iowa
School of Business
Cedar Falls, IA 50613

UNITED STATES—KANSAS

1691
Appropriate Technology Group
Route 1
Box 93-A
Oskaloosa, KS 66066

1692
Windustries, Inc.
2429 A Rosebud Lane
Lawrence, KS

1693
Kansas City Regional Development

Association
P.O. Box 912
Shawnee Mission, KS 66201

1694
Kansas State University
Food & Feed Grain Institute
Shellenberger Hall
Manhattan, KS 66506

1695
Seaton International Trade Corporation
801 Wisconsin Street
Neodesha, KS 66757

UNITED STATES—LOUISIANA

1696
American Sugarcane League
416 Whitney Building
New Orleans, LA 70130

1697
Rice Experiment Station
Crowley, LA

1698
St. Gabriel Experiment Station
Box 34
St. Gabriel, LA

1699
Southern Regional Research Laboratory
c/o Chief of the Oilseed Crops Laboratory
1100 Robert E. Lee Blvd.
New Orleans, LA

UNITED STATES—MAINE

1700
Gorham International, Inc.
Gorham, ME 04038

1701
The Grist Mill
90 Depot Road
Eliot, ME 03903

1702
Johnny's Selected Seeds
Albion, ME 04910

1703
Maine Audubon Society
118 Rt. 1
Falmouth, ME 04105

1704
Maine Organic Farmers and Gardeners Association
RFD 2, Bump Hill Road
Union, ME 04642

1705
Northeast Carry
P.O. Box 187
Hallawell, ME 04347

1706
Shelter Institute
72 Front Street
Bath, ME 04530

1707
The Small Farm Research Association
Greenwood Farm
Harborside, ME 04642

1708
Solar Wind Company
P.O. Box 7
East Holden, ME 04429

1709
TRANET
P.O. Box 567
Rangley, ME 04971

1710
Zephyr Wind Dynamo Company
P.O. Box 241
Brunswick, ME 04011

UNITED STATES—MARYLAND

1711
Foundation for Self-Sufficiency, Inc.
Research Center
35 Maple Avenue
Catonsville, MD 21228

1712
Inter-Culture Associates, Inc.
1800 Drexel Street
Hyattsville, MD 20783

1713
International Compendium
Division of Solar Science Industries
10762 Tucker Street
Beltsville, MD 20705

1714
International Program for Human Resources Development
P.O. Box 30216
Bethesda, MD 20014

1715
International Solar Energy Society
12441 Parklawn Dr.
Rockville, MD 20852

1716
International Sugar Research Foundation
7316 Wisconsin Ave.
Bethesda, MD 20014

1717
National Solar Heating and Cooling Information Center
P.O. Box 1607
Rockville, MD 20850

1718
School of Living
Heathcote Center
Route 1, Box 129
Freeland, MD 21053

1719
University of Maryland
Cooperative Extension Service
College Park, MD 20742

1720
Volunteers in Technical Assistance (VITA)
3706 Rhode Island Avenue
Mt. Rainier, MD 20822

1721
World Environmental Directory
Business Publishers, Inc.
P.O. Box 1067
Silver Spring, MD 20910

1722
World Water Resources
7315 Wisconsin Ave.
Bethesda, MD 20014

UNITED STATES—MASSACHUSETTS

1723
ACCION/AITEC
10-C Mount Auburn St.
Cambridge, MA 02138

1724
Biodynamic Farming and Gardening Association
165 West St.
Duxbury, MA 02543

1725
The Bolton Institute
4 Linden Square
Wellesley, MA 02181

1726
Boston Architecture Center
Service for Energy Conservation in Architecture
320 Newbury Street
Boston, MA 02115

1727
Boston Wind
2 Mason Court
Charlestown, MA 02129

1728
Center for Community Economic Development
639 Massachusetts Ave., Suite 316
Cambridge, MA 02139

1729
Centre for the Study of Development and Social Change
1430 Massachusetts Ave.
Cambridge, MA 02138

1730
Circuit Engineering
15 Ellis Road
Weston, MA 02193

1731
Community Training Resources
12 Maple Ave.
Cambridge, MA 02139

1732
Dollars and Sense
324 Somerville Ave.
Somerville, MA 02143

1733
Educational Exchange of Greater Boston, Inc.
17 Dunster Street
Cambridge, MA 02138

1734
Hippocrates Health Institute

25 Exeter Street
Boston, MA 02116

1735
International Independence Institute
Box 183 West Road
Ashby, MA 01431

1736
Wyman E. Kilgore
1520 Bailey Street
Hastings, MA 55033

1737
Chris Logan
8 Rodney Rd.
Bedford, MA 01730

1738
Nacul Environmental Design Center
592 Main Street
Amherst, MA 01002

1739
The New Alchemy Institute
Box 432
Woods Hole, MA 02543

1740
New England Appropriate Technology Network (NEAT-NET)
Box 134 Harvard Sq.
Cambridge, MA 02138

1741
New Roots Magazine
Box 459
Amherst, MA 01002

1742
Office to Coordinate Energy Research & Education
A-25 Graduate Research
University of Massachusetts
Amherst, MA 01002

1743
Oxfam America
302 Columbus Ave.
Boston, MA 02116

1744
SESPA
Scientists and Engineers for Social and Political Action
9 Walden Street
Jamaica Plain, MA 02130

1745
Technical Development Corp.
11 Beacon St.
Boston, MA 02108

1746
Whitewood Stamps, Inc.
61 Chapel Street
Newton, MA 02158

1747
Williams College
Center for Development Economics
Williamstown, MA 01267

UNITED STATES—MICHIGAN

1748
Alger-Marquette Community Action Board
600 Altamont
Marquette, MI 49855

1749
Environmental Energies, Inc.
Box 73, Front Street
Copernish, MI 49625

1750
Environmental Research Institute of Michigan
Post Office Box 618
Ann Arbor, MI 48107

1751
David Hartman
1337 Wilmot, #9
Ann Arbor, MI 48104

1752
Michigan Organic Growers and Buyers Association
Box 136
Decatur, MI 49045

1753
Sunstructures, Inc.
201 E. Liberty Street
Ann Arbor, MI 49240

1754
U-Form Systems and Technology, Inc.
29200 Vassar Ave., Suite 700
Livonia, MI 48152

1755
Upland Hills Ecological Awareness

Center
2575 Indian Lake Road
Oxford, MI 48051

1756
Western Michigan University
Department of Political Science
Kalamazoo, MI 49008

1757
Doc Whiting
328 John Street
Ann Arbor, MI 48104

UNITED STATES—MINNESOTA

1758
Alternative Industries Association
Route 2, Box 90-A
Milaca, MN 56353

1759
Alternative Sources of Energy, Inc.
Route 2, Box 90-A
Milaca, MN 56353

1760
American Association of Cereal Chemists
3340 Pilot Knob Road
St. Paul, MN 55121

1761
The Big Outdoors People
2201 N.E. Kennedy St.
Minneapolis, MN 55413

1762
Center for Local Self Reliance
3302 Chicago Ave.
So. Minneapolis, MN 55407

1763
Community Design Center of Minnesota
118 East 26th St.
Minneapolis, MN 55408

1764
Energy Research and Development Corporation
3544 Emerson Avenue South
Minneapolis, MN 55408

1765
Jopp Electrical Works
Princeton, MN 55371

1766
Minnesota Organic Growers and Buyers Association
624 Jefferson St., N.E.
Minneapolis, MN 55413

1767
Rutan Research
12489 Norell Avenue, N.
Stillwater, MN 55082

1768
University of Minnesota
Project Ourobors
School of Architecture and Landscape Architecture
110 Architecture Bldg.
Minneapolis, MN 55455

UNITED STATES—MISSISSIPPI

1769
Mississippi Research and Development Center
Post Office Drawer 2470
Jackson, MS 39205

1770
Mississippi State University
Engineering and Industrial Research Station
Mississippi State, MS 39762

1771
Solvent Extraction Plant
Planters Manufacturing Co.
Clarksdale, MS 38614

UNITED STATES—MISSOURI

1772
Acres, U.S.A.
10227 E. 61st Street
Raytown, MO 64133

1773
Washington University
Department of Technology and Human Affairs, and the Center for Development Technology
Box 1106
St. Louis, MO 63130

UNITED STATES—MONTANA

1774
Alternative Energy Resources Organization (AERO)
435 Stapleton Bldg.
Billings, MT 59101

1775
Montana Department of Natural Resources and Conservation
Energy Planning Division
32 South Ewing
Helena, MT 59601

1776
Montana Energy and Magnetohydrodynamics Research and Development Institute (MERDI)
Butte, MT

1777
Montana Trust for People and Land
P.O. Box 12
Helena, MT 59601

1778
National Center for Appropriate Technology
P.O. Box 3838
Butte, MT 59701

1779
Sunshine Engineering
P.O. Box 1776
Gildford, MT 59525

UNITED STATES—NEBRASKA

1780
Center for Rural Affairs
P.O. Box 405
Walthill, NE 68067

1781
Midwest Organic Producers Association
c/o Center for Rural Affairs
P.O. Box 405
Walthill, NE 68067

1782
Nebraska's New Land Review
Center for Rural Affairs
P.O. Box 405
Walthill, NE 68607

1783
Small Farm Energy Project
Center for Rural Affairs
P.O. Box 736
Hartington, NE 68739

UNITED STATES—NEVADA

1784
Desert Research Institute
University of Nevada System
Reno, NV 89507

UNITED STATES—NEW HAMPSHIRE

1785
Donovan and Bliss
Chocorua, NH 03817

1786
High Mowing Farm Project
High Mowing School
Wilton, NH 03086

1787
Solar Survival
Cherry Hill Road
Harrisville, NH 03450

1788
Solid Waste Recovery Company
16 Meserve Rd.
Durham, NH 03824

1789
Total Environment Action
Church Hill
Harrisville, NH 03450

1790
University of New Hampshire
Cooperative Extension Service
Pettee Hall
Durham, NH 03824

1791
University of New Hampshire
Extension Consumer Information
Taylor Hall
Durham, NH 03824

UNITED STATES—NEW JERSEY

1792
Calmac Manufacturing Corp.

150 South Van Brunt Street
Englewood, NJ 07631

1793
Cook College
Department of Biological and Agricultural Engineering
P.O. Box 231
New Brunswick, NJ 08903

1794
Public Service Electric and Gas Company
Research and Development Dept.
80 Park Place
Newark, NJ 07101

UNITED STATES—NEW MEXICO

1795
Adobe News, Inc.
P.O. Box 702
Los Lunas, NM 87031

1796
Appropriate Technology Research
1938 Hano Road
Santa Fe, NM 87501

1797
La Cooperacion del Pueblo
Box 96
Tierra Amarilla, NM 87575

1798
Council for Sustainable Growth in New Mexico
634 Garcia Street
Santa Fe, NM 87501

1799
Foundation for Rural Technology
P.O. Box 8
Embudo, NM 87531

1800
Integrated Living Systems
P.O. Box 537
Tijeras, NM 87059

1801
New Mexico Division of Human Resources
ATTN: Energy Program Coordinator
Villagra Bldg., Room 119
Santa Fe, NM 87503

1802
New Mexico Organic Growers Association
1312 Lobo Place, N.E.
Albuquerque, NM 81706

1803
New Mexico Solar Energy Association
P.O. Box 2004
Santa Fe, NM 87501

1804
New Mexico State University
Box 3450
Las Cruces, NM 88003

1805
Potrero Biotechnics
El Rito, NM 87530

1806
Sandia Laboratories
Solar Total Energy Program
Albuquerque, NM 87115

1807
Solar Sustenance Project
Rt. 1, Box 107AA
Santa Fe, NM 87501

1808
University of California
Los Alamos Scientific Laboratories
P.O. Box 1663
Los Alamos, NM 87545

1809
University of New Mexico
Department of Mechanical Engineering
Albuquerque, NM 87131

1810
University of New Mexico
Technology Application Center
Albuquerque, NM 87131

1811
The Walden Foundation
P.O. Box 5
El Rito, NM 87530

1812
Zomeworks
P.O. Box 712
Albuquerque, NM 87130

UNITED STATES—NEW YORK

1813
Agricultural Development Council
1290 Avenue of the Americas
New York, NY 10019

1814
Allahabad Agricultural Institute, Inc.
M.O. Centre
Stony Point, NY

1815
American Council of Voluntary Agencies for Foreign Serivice, Inc.
Technical Assistance Information Clearing House
200 Park Avenue South
New York, NY 10003

1816
Applied Forestry Research Institute
Tropical Wood Information Center
Syracuse Campus
Syracuse, NY 13210

1817
ARCA Foundation
100 East 85th Street
New York, NY 10028

1818
Association for the Study of Man—Environment Relations
P.O. Box 57
Orangeburg, NY 10962

1819
Bio-Energy Systems, Inc.
P.O. Box 375
Spring Glen, NY 12483

1820
Brookhaven National Laboratory
Upton, NY 11973

1821
CARE
600 First Avenue
New York, NY 10016

1822
Catholic Foreign Missions
Maryknoll, NY 10543

1823
Church World Service
475 Riverside Drive
New York, NY 10027

1824
Committee for Social Responsibility in Engineering
475 Riverside Drive
New York, NY 10017

1825
Congregational Christian Service Committee
475 Riverside Drive
New York, NY 10027

1826
Construction Systems Development
Albany Post Road
Croton-on-Hudson, NY 10520

1827
Consumer Action Now
49 East 53rd St., 9th Fl.
New York, NY 10022

1828
Cornell University
Office of Energy Programs
Ithaca, NY 14853

1829
Cornell University
Entomology and Agricultural Science
Ithaca, NY 14853

1830
Cornell University
Program on Politics for Science and Technology in Developing Nations
180 Uris Hall
Ithaca, NY 14853

1831
Corporate Information Center
Room 566
475 Riverside Drive
New York, NY 10027

1832
Dawes Hill Commune
P.O. Box 53
West Danby, NY 14896

1833
Emerging Professional International Conference
313 W. 87th St.
New York, NY 10024

1834
Ford Foundation

International Division
320 East 43rd Street
New York, NY 10017

1835
Foundation Center
888 Seventh Ave.
New York, NY 10019

1836
Friends World College
Lloyd Harbor
Huntington, NY 11743

1837
Garden Way Manufacturing Company
102nd and 9th Avenue
Troy, NY 12180

1838
Institute on Man and Science
Rensselaerville, NY 12147

1839
International Agricultural Development Service
1133 Avenue of the Americas
New York, NY 10036

1840
International Education Development, Inc.
924 West End Avenue
New York, NY 10025

1841
International Executive Service Corps.
545 Madison Avenue
New York, NY 10022

1842
International Institute of Rural Reconstruction
1775 Broadway
New York, NY 10570

1843
Hotline International
Glen Leet
54 Riverside Drive
New York, NY 10024

1844
Marathon Heater Company, Inc.
Box 165, R.D. 2
Marathon, NY 13803

1845
Martha Stuart Communications, Inc.
66 Bank Street
New York, NY 10014

1846
Mohawk Valley Community College
Civil Technology Department
1101 Sherman Drive
Utica, NY 13501

1847
Nassau Division of Energy Resources
240 Old Country Rd.
Mineola, NY 11501

1848
PACT—Private Agencies Collaborating Together
777 United Nations Plaza
New York, NY 10017

1849
Rockefeller Brothers Fund
30 Rockefeller Plaza
New York, NY 10020

1850
Solarvision, Inc.
Box Z
Port Jervis, NY 12771

1851
State University of New York
The Institute for Man and Environment
Plattsburgh, NY 12901

1852
Richard G. Stein and Associates
Architects
588 Fifth Avenue
New York, NY 10036

1853
Technical Assistance Information Clearinghouse (TAICH)
200 Park Avenue South
11th Floor
New York, NY 10003

1854
Tropical Wood Information Center
Applied Forestry Research Institute
College of Forestry
Syracuse Campus
Syracuse, NY 13210

1855
UNDP, Division of Information
DC Bldg., Rm. 1912

1 U.N. Plaza
New York, NY 10017

1856
United Nations
Water Resources Branch
Center for Natural Resources, Energy, and Transport
New York, NY 10017

1857
United Nations Children's Fund (UNICEF)
866, United Nations Plaza
New York, NY 10017

1858
World Crafts Council
Office for Crafts Development
29 West 53rd Street
New York, NY 10019

1859
World Education
1414 Sixth Avenue
New York, NY 10019

UNITED STATES—NORTH CAROLINA

1860
The Celo Community, Inc.
R.R. #5
Burnsville, NC 28714

1861
Lorien House
P.O. Box 1112
Black Mountain, NC 28711

1862
North Carolina State University
Agricultural Extension Service
Raleigh, NC 27607

1863
Rural Advancement Fund
Frank Porter Graham Experimental Farm and Training Center
2128 Commonwealth Ave.
Charlotte, NC 28205

1864
Solar Products Information and Engineering
P.O. Box 506
Columbus, NC 28722

1865
Sunshine Engineering
P.O. Box 61
Westfield, NC 27053

1866
Sunspace, Inc.
Box 71A, Rt. 5
Burnsville, NC 28714

1867
University of North Carolina
Department of Environmental Sciences and Engineering
School of Public Health
Chapel Hill, NC 27514

UNITED STATES—NORTH DAKOTA

1868
Minor Oilseeds Research Unit
Plant Pathology Department
Fargo, ND 58102

UNITED STATES—OHIO

1869
Anderson IBEC
19699 Progress Drive
Strongsville, OH 44136

1870
Appalachian Rural Alternatives Network
P.O. Box 135
Bainbridge, OH 45612

1871
Battelle Memorial Institute
505 King Avenue
Columbus, OH 43201

1872
Community Service, Inc.
Box 243
Yellow Springs, OH 45387

1873
Creative Living
472 W. 8th Ave.
Columbus, OH 43210

1874
French Oil Mill Machinery Co.
1099 Greene Street
Piqua, OH 45356

1875
Griscom Morgan
Community Services, Inc.

Whitemen St.
Yellow Springs, OH 45387

1876
The James Leffel & Co.
426 East Street
Springfield, OH 45501

1877
Ohio University
Center for International Studies
Athens, OH 45701

1878
Raven Rocks, Inc.
Route 1
Beallsville, OH 43716

1879
Sunpower, Inc.
48 West Union St.
Athens, OH 45701

1880
University of Dayton
Strategies for Responsible Development
Dayton, OH 45469

1881
Western Reserve Foundation
P.O. Box 1412
Hudson, OH 44236

UNITED STATES—OKLAHOMA

1882
Oklahoma State University
Energy Laboratory
School of Electrical Engineering
Stillwater, OK 74074

1883
World Neighbors
5116 N. Portland Avenue
Oklahoma City, OK 73112

UNITED STATES—OREGON

1884
Amity Foundation
1820 Olive
Eugene, OR 97401

1885
Anchor Tools and Woodstoves
618 N.W. Davis
Portland, OR 97209

1886
Flexible Ways to Work
2683 Alder
Eugene, OR 97405

1887
Full Circle
760 Vista Avenue, S.E.
Salem, OR 97302

1888
Nutrition Information Center
239 S.E. 13th Ave.
Portland, OR 97214

1889
Oregon State University
International Plant Protection Center
Corvallis, OR 97331

1890
People's Transhare, inc.
P.O. Box 8393
Portland, OR 97207

1891
Portland State University
Systems Science Department
P.O. Box 751
Portland, OR 97207

1892
RAIN
2270 N.W. Irving
Portland, OR 97210

1893
Service Exchange
3543 S.E. Main
Portland, OR 97214

1894
University of Oregon
Action Now/ASUO EMU II
Eugene, OR 97403

1895
University of Oregon
Department of Architecture
Eugene, OR 97403

UNITED STATES—PENNSYLVANIA

1896
American Friends Service Committee
1501 Cherry St.
Philadelphia, PA 19102

1897
Approtech, Inc.
1700 Meadow Road
Southampton, PA 18966

1898
Aquarian Research Foundation
Box P-4120
5620 Morton Street
Philadelphia, PA 19144

1899
Bio-Utility Systems, Inc.
P.O. Box 135
Narberth, PA 19072

1900
Bryn Gweled Homesteads
1100 Woods Rd.
Southampton, PA 18966

1901
Carnegie-Mellon University
Advanced Building Studies Program
CFA 319
6000 Forbes Avenue
Pittsburgh, PA 15213

1902
Clean Energy Systems
RD 1, Box 366
Elysburgh, PA 17824

1903
Franklin Research Center
20th and Race Streets
Philadelphia, PA 19103

1904
R. Buckminister Fuller and Associates
3500 Market St.
Philadelphia, PA 19104

1905
Hydroheat Division
Ridgway Steel Fabricators, Inc.
Ridgway, PA 15853

1906
Mennonite Economic Development Associates
21 South 12th Street
Akron, PA 17501

1907
Opportunities Industrialization Centers International (OICI)
140 West Tulpenhocken Street
Philadelphia, PA 19144

1908
Commonwealth of Pennsylvania
Bureau of Human Resources
Department of Community Affairs
Harrisburgh, PA 17120

1909
Philadelphia Bicycle Coalition
3410 Baring Street
Philadelphia, PA 19104

1910
Project America 1976
4300 Freemansburg Ave.
Hopeville (Easton), PA 18042

1911
Ridgway Steel Fabricators, Inc.
Hydroheat Division
Ridgway, PA 15853

1912
Rodale Press, Inc.
33 East Minor Street
Emmaus, PA 18049

1913
University of Pittsburgh
Project Solo
Computer Science Department
311 Alumni Hall
Pittsburgh, PA 15260

UNITED STATES—RHODE ISLAND

1914
REDE Corporation
P.O. Box 212
Providence, RI 02901

UNITED STATES—SOUTH CAROLINA

1915
Rural Housing Research Unit
Agricultural Research Service, USDA
P.O. Box 792
Clemson, SC 29631

UNITED STATES—SOUTH DAKOTA

1916
Center for Community Organization and Area Development
2118 South Summit Ave.
Sioux Falls, SD 57105

UNITED STATES—TENNESSEE

1917
The Farm
156 Drakes Lane
Summertown, TN 38483

1918
Institute for Energy Analysis
P.O. Box 117
Oak Ridge, TN 37830

1919
National Cottonseed Products Association
P.O. Box 12032
Memphis, TN 38112

1920
Oak Ridge National Laboratories
Oak Ridge, TN 37830

UNITED STATES—TEXAS

1921
Agricultural Research and Extension Center at Beaumont
Route 5, Box 784
Beaumont, TX 77706

1922
East Texas Council of Governments
5th Floor, Allied Citizens Bank Bldg.
Kilgore, TX 75662

1923
Food Protein Research and Development Center
Faculty Mail Box 63
College Station, TX 77843

1924
Ideation
5435 Ellsworth
Dallas, TX 75206

1925
Institute of Food Science and Engineering
College of Agriculture
College Station, TX 77843

1926
Intertect
P.O. Box 10502
Dallas, TX 75207

1927
H.T. McGill and Company
903 Fairoaks
Houston, TX 77001

1928
Oil Mills
Anderson Clayton and Company
Tenneco Building
1010 Milam
Houston, TX

1929
Over the Garden Fence
3960 Cobblestone Dr.
Dallas, TX 75229

1930
Rice Council for Market Development
3917 Richmond Ave.
Houston, TX 77027

1931
Texas A&M University
Texas Intensified Farm Planning Program
Texas Agricultural Extension Service
College Station, TX 77840

1932
Texas A&M University
Oilseed Products Division
Food Protein Research and Development Center
College Station, TX 77843

1933
University of Texas
Center for Maximum Potential Building Systems
School of Architecture
Austin, TX 78712

UNITED STATES—UTAH

1934
Institute of Maintenance Research
2217 Evergreen Ave.
Salt Lake City, UT 84109

1935
Utah State University
Consortium for International Development
UMC 35
Logan, UT 84322

1936
Utah Valley Hospital
Department of Rural Health
1034 North 500 West
Provo, UT 84601

UNITED STATES—VERMONT

1937
Appropriate Technology Corporation
P.O. Box 121
Townshend, VT 05353

1938
Enertech Corporation
P.O. Box 420
Norwich, VT 05055

1939
Garden Way Laboratories
Box 97
Charlotte, VT 05445

1940
Gardens for All, Inc.
Bay and Harbor Roads
Box 371
Shelburne, VT 05482

1941
Grassy Brook Village, Inc.
RFD 1, Box 39
Newfane, VT 05345

1942
Natural Organic Farmers Association
Box 247
Plainsfield, VT 05667

1943
North Wind Power Co., Inc.
P.O. Box 315
Warren, VT 05674

1944
Vermont Tomorrow
5 State Street
Montpelier, VT 05602

1945
Wood Energy Institute
Box 1, Fiddlers Green
Waitsfield, VT 05673

UNITED STATES—VIRGINIA

1946
Alternative Sources of Energy
928 2nd Street, S.W.
Apt. 4
Roanoke, VA 24016

1947
Institute for International Development, Inc.
8150 Leesburg Pike, Suite 504
Vienna, VA 22180

1948
National Recreation and Park Association
Park Project on Energy Interpretation
1601 N. Kent
Arlington, VA 22209

1949
Twin Oaks
Louisa, VA 23093

1950
U.S. Department of Commerce
National Technical Information Service (NTIS)
5285 Port Royal Road
Springfield, VA 22161

UNITED STATES—WASHINGTON

1951
Alternative Energy Systems
P.O. Box 295
Winthrop, WA 98862

1952
Bridge Design Cooperative
General Delivery
Burton, WA 98013

1953
Ecotope Group
747 16th Street East
Seattle, WA 98112

1954
Environmental Farm Project
c/o ESD 110
Shoreline Schools Administration, N.E.
158th and 20th Ave., N.E.
Seattle, WA 98155

1955
Metastasis

P.O. Box 128
Marblemount, WA 98267

1956
New World Computer Services
P.O. Box 5415
Seattle, WA 98105

1957
Northwest Organic Food Producers Association
Route 2, Box 2152
Toppenish, WA 98948

1958
Northwest Regional Foundation (NRF)
P.O. Box 5296
Spokane, WA 99205

1959
Program of Social Management of Technology
316 Guggenheim, FS-15
University of Washington
Seattle, WA

1960
Small Hydroelectric Systems and Equipment
P.O. Box 124
Custer, WA 98240

1961
Small Towns Institute
Box 517
Ellensburg, WA 98926

1962
TILTH
Rt. 2, Box 190-A
Arlington, WA 98223

UNITED STATES—WEST VIRGINIA

1963
Goodheart Farm
Rt. 2, Box 206
Berkeley Springs, WV 25411

1964
Solar Age Press
Indian Mills, WV 24949

1965
West Virginia Department of Agriculture
Small Farm Project—Produce Development Section
State Capitol Building
Charleston, WV 25305

1966
West Virginia University
Department of Chemical Engineering
Morgantown, WV 26506

UNITED STATES—WISCONSIN

1967
ADVOCAP Inc. Community Action Program
174 W. Division St.
Fond du Lac, WI 54935

1968
Beloit College
World Affairs Center
Beloit, WI 53511

1969
Countryside
312 Portland Rd.
Highway 19 East
Waterloo, WI 53594

1970
Energy Research, Development and Applications Project
Forest Products Laboratory
Madison, WI 53705

1971
Fabric Appliance Company
RR 1, Box 150 A
Baldwin, WI 54002

1972
Organic Growers and Buyers Association
Winding Road Farm
Boyceville, WI 54724

1973
University of Wisconsin
International Cooperative Training Center
Madison, WI 53705

1974
University of Wisconsin-Madison
College of Engineering
Madison, WI 53706

1975
University of Wisconsin

Madison Center for Health Service
610 North Walnut St.
Madison, WI 53706

UNITED STATES—WYOMING

1976
Community Action of Laramie County, Inc.
Bell Bldg., Suite 400
Cheyenne, WY 82001

1977
Wyoming Specialties, Inc.
Box 721
Gillette, WY 82716

INDEX